DEMOCRACY AND MATHEMATICS EDUCATION

In *Democracy and Mathematics Education*, Kurt Stemhagen and Catherine Henney develop a way of thinking about the nature and purposes of math that is inclusive, participatory, and thoroughly human. They use these ideas to create a school mathematics experience that can enhance students' math abilities and democratic potential. They locate mathematics' origins in human activity and highlight the rich but often overlooked links between mathematical activity and democratic, social practices. Democratic mathematics education foregrounds student inquiry and brings to light the moral dimensions of a discipline that has both remarkable utility and inevitable limitations. For math educators, the book's humanities approach helps to see the subject anew. For philosophers, it provides an important real world context for wrestling with perennial and timely questions, engaging democratic and evolutionary theory to transform school math. This alternative approach to mathematics and mathematics education provides a guide for how to use math to make democracy a larger part of school and wider social life.

Kurt Stemhagen is Professor of Education at Virginia Commonwealth University. He earned his Ph.D. in Social Foundations of Education with a concentration in Philosophy of Education from the University of Virginia. At VCU, he teaches doctoral courses on philosophy of educational research and philosophy of education as well as undergraduate and master's courses on democracy, diversity, and ethics. Dr. Stemhagen is co-founder and an active member of Richmond Teachers for Social Justice, a group dedicated to creating a just, democratic, sustainable, and caring society through education, solidarity, and social action.

Catherine Henney is a veteran math educator, having worked with students and teachers as both a grade level teacher and a K-8 mathematics specialist. She also teaches mathematics education courses to pre-service and in-service teachers in Virginia. She is co-author of the book *It's Elementary: A Parent's Guide to K-5 Mathematics*, published by the National Council of Teachers of Mathematics, and is currently a doctoral student at Virginia Commonwealth University.

DEMOCRACY AND MATHEMATICS EDUCATION

Rethinking School Math for Our Troubled Times

Kurt Stemhagen and Catherine Henney

NEW YORK AND LONDON

First published 2021
by Routledge
52 Vanderbilt Avenue, New York, NY 10017

and by Routledge
2 Park Square, Milton Park, Abingdon, Oxon OX14 4RN

Routledge is an imprint of the Taylor & Francis Group, an informa business

Library of Congress Cataloging-in-Publication Data
A catalog record for this title has been requested

ISBN: 978-0-367-60821-7 (hbk)
ISBN: 978-0-367-60820-0 (pbk)
ISBN: 978-1-003-10063-8 (ebk)

Typeset in Bembo
by Taylor & Francis Books

CONTENTS

ILLUSTRATIONS

Figures

Tables

ACKNOWLEDGEMENTS

We wish to acknowledge the contributions of cherished family members, colleagues, and friends. Without their encouragement and assistance, this book might never have been realized. We are indebted to Kristin Stemhagen, Bronwynn Gabriel, and Emma Shachat for their expertise and help in finalizing the manuscript. To our respective mentors, Eric Bredo and Joy Whitenack, we wish to express our gratitude for your guidance in the ways of philosophy, writing, and life. Joy, your words seem to swell with faith in children's brilliance. Eric, you have demonstrated how much intellectual work can be done while hiking together on a mountain trail, or walking and talking through city streets. We are indebted to you both.

FOREWORD

The Problem: Why Teach Math

The Traditional Responses

There are two common default positions to the question why study math. The first is that math is a requirement for entrance into an elite university and a high paying position. The second is that math is a critical instrument for the advancement of the national interest. Both of these are obviously true, but both are essentially empty, raising more questions than they answer. Yes, most students—legacies excepted—complete high level math to get into most elite universities, but what kind of math education *should* be considered appropriate for an educated person? And yes, math is certainly important for achieving the national interest, but then what kind of interest should a *democratic* nation have and how can math best serve it? One response to the question why study math is largely missing—to advance democracy.

The Tension Between Democratic Education and Math Education

Mathematics and democracy are not considered good traveling companions, and of all the subjects taught in school mathematics are seen to be the least amenable to democratic norms. Good social studies teachers encourage debate and discussion; good literature teachers promote unique and insightful interpretations; good science teachers nurture close observation of nature and careful experimentation. But "good" math teachers—well, they want right answers and high scores on standardized tests.

In addition, math has a reputation for separating the weak lambs from the powerful lions—hardly a democratic image. In many math classrooms the same few students always have their hands up, ready with the correct answer to the toughest problem while the others hide—heads down, staring at their desks—hoping not to be called on as they belatedly struggle to understand the answer to the last question. In the teacher's lounge there is the buzz about the math whiz whether he—and it has usually been a "he"—will choose MIT or Caltech. And there is the sad collective shrug expressing pity for that student who is fated to "never really get it." Students are marked by the kind of math they take—smart kids take advanced trig and calculus and slow ones take general or consumer math. And these markers have wider consequences in terms of the attention students receive, the clubs they belong to, and the kinds of friends they have.

Many people accept this as the way things are and must forever be, and then a few work to meliorate the side effects. If the very nature of math is undemocratic, then the solution is not to change math, but to address the indirect, undemocratic side effects of the exclusivity and elitism that follow in its wake. Open up the clubs, encourage wider friendships, and find ways to mix students everywhere else—but not in trig or calculus. From this standpoint math itself is cut and dried. 2+2 equals 4. Always has, always will. No voting will change that fact. No debate is necessary; no amount of discussion will make 2+2=5.

Making Math Education Democratic

Democracy and Mathematics Education is a needed response to the view that math is inherently undemocratic. By tracing mathematics to its roots in human activity the authors expose the deep, but forgotten, connection between mathematical knowledge and democratic, communal practices. One of the aims of math education should be to expose the lost connection between math and democracy. Hence their conception of democratic math education (DME) supports learning activities that will enable students to understand how the tool of math serves to open up possibilities for democracy—not to close them. With DME math is demystified and students become aware of some of the ways it can aid good decision making. In the process students come to understand math not as the cut and dried subject it is often presented as, but rather as part of an evolutionary process that changes in response to human needs and interests.

How Math Mystifies Experience

Anyone who has had to rank order something can get a sense of the authors' concern. My experience on the admissions committee to the Medical Scholars program at the University of Illinois in the early 1980s can serve as an example. On the surface the process looks very straightforward. It seems to begin when all the members of the admission committee are asked to rank order the candidates by certain relevant criteria—say scores on the MCAT, the courses the student took, the grades received, the quality of the undergraduate college, extracurricular activity, student background, other relevant experience, the impressions of their interviews with faculty. Each member of the committee then provides a score on each of these categories, adds up the individual items, and then reports the total to the chair of the committee who then adds them all together and averages them to provide the rank order.

The process certainly seems very straightforward and objective. Who can argue with the numbers—after all the scores have been added up and averaged Joe had a total of 80 while Jack had only a total of 79. Admit Joe, reject Jack! The process then is represented as fair and objective. However, the numbers can conceal some real messiness. Some members of the committee believed that the MCAT score was worthless when it came to predicting the quality and dedication of future doctors. Some members of the committee frowned on any course that was not in the "hard sciences." Others believed that a well-rounded liberal education was the best preparation for a life as a doctor, not the narrow pre-medical program. As for the reputation of the undergraduate program, everyone allowed that most of that was hearsay—although everyone used it. And the faculty impression—scored on another numerical scale—depended a great deal on the chemistry between the faculty member and the candidate.

And then there were the less visible components of this "objective" process. All of the members of the committee were White and all were male. Not surprisingly at that time—the early 1980s—the vast majority of applicants were also White men. Of course, by the time we got to adding the numbers up, and delivering our verdict (Joe in; Jack out!) we rarely had any problem with the addition. The problem was not in the math itself, but in the fact that the math made everything appear so objective and beyond question.

In fact, the process itself was political in a deep way, but the numbers obscured this fact. At the time few thought to ask where all the Black or female candidates were. The simplest answer was, of course, that they did not choose to apply. A more complicated one was that they did not have the math and science prerequisites needed to apply. To then have asked why not would have opened up the very messy question of why some groups of students choose the subjects they do. And, then on to the further question: what counts as a real choice? All of these perfectly reasonable questions were hidden beneath the apparent objectivity of the numbers. In other words, math can lend legitimacy to decisions that actually need further investigation. An important aim of DME is to teach students to be sensitive to this and open up inquiry that might otherwise be closed and to see math as a tool for expanding possibilities. Part of this is to learn how to identify the inappropriate use of math to mystify the rationale for a decision. But there is more to the process than simply addressing political issues. Given the right kinds of problems students can begin to see just how math can be used to aid choice and creativity.

It would be misleading to think of DME as simply another combatant in the infamous math wars. It is consistent with all and any approaches to the teaching of math as long as certain conditions are maintained. One of these conditions that the authors pay particular attention to is the need to avoid reifying math in a way that removes it from the sphere of human need and interest, a condition that is of considerable concern in traditional classrooms with the emphasis on well-worn algorithms and the correct answer. Two additional conditions are also worth mentioning. Philosophers call these conditions side constraints.

Democracy and Side Constraints

A side constraint sets the boundary conditions of a practice. It is not the goal of the practice itself but rather expresses a boundary that a practice needs to avoid in pursuing its goal. A side constraint is like the speed limit that you are expected to obey as you drive, late as you are, to your appointment with the doctor. It is not your goal—that is to get to the doctor—but it is something that needs to be followed in meeting your goal.

The two side constraints that I would like to mention here are: 1. Avoid the Lysenko error and 2. Don't make a fetish of authenticity. The Lysenko error is named after the Soviet agronomist who is blamed for single-handedly destroying the Russian harvest because he rejected Mendelian genetics in favor of pseudoscientific ideas that conformed to the Soviet ideology. This side constraint warns against distorting reasonable findings for the sake of some preexisting idea, often political, about the most desirable outcome. To ignore this constraint can lead to the dogmatic rejection of dogmatism, or what my colleague Eric Bredo refers to as dogmatic anti-dogmatism. The fetish of authenticity is a warning about the readiness to accept any response as long as it is seen as the expression of a "really true self."

As I mentioned, it would be a mistake to view DME as simply another combatant in the long-standing math wars. Rather, it is consistent with math at the highest, most rigorous level, where content is critical. And it is also consistent with student-centered math programs, where the student's self-esteem is more valued than getting the right answer. And it is consistent with social justice math education that in a very different way is concerned with using math for democratic purposes. To the extent that democracy is an issue with these different programs it is most likely to be with the extent to which these side constraints are honored, but this is an issue for a different time.

By demonstrating how math and democracy are interrelated, Stemhagen and Henney have performed a valuable service for both philosophers and math teachers. Philosophers can see how ideas guide real-life teaching and teachers can get a better idea of how philosophy can help them think through their own practice. In the end of course, *Democracy and Mathematics Education* is about democracy and how to use math to make it a larger part of school life.

Walter Feinberg

PART I
The Problem

1

INTRODUCTION

Rethinking Math for Our Troubled Times

Mathematics is a curious discipline. Each of us has a math history and, in all likelihood, a complicated relationship to the subject. Our memories of the math classroom may tell a story of success and delight. Or they may tell a story of frustration, disinterest, or even dread. No other school subject is nearly so divisive. At the mere mention of mathematics, we quickly sift and sort ourselves into categories internalized long ago, probably during our childhood.

How is it that mathematics can mean so much to some and so little to others? Why such widely disparate experiences? And they are indeed wide and disparate. Unequal.

This is by no means a new phenomenon. Historically, math has been cordoned off from the masses and reserved only for the college-bound, particularly for those advancing to quantitative fields such as engineering. This "lack of welcome," as Francis Su (2020) has described it, has far-reaching injurious social consequences. It is not simply that an exclusionary mathematics is undemocratic; it also works to forestall the flourishing of democracy itself.

While we are not claiming that a non-democratic school mathematics experience is *the* reason that our democracy is in its current state, at the very least we do see math class as a missed opportunity. We have multiple reasons for writing this book and developing a democratic philosophy of mathematics education and a related set of teaching/learning practices. When one of the authors first started with this line of thinking in the early 2000s, its purpose was primarily to improve school mathematics. Over time, we have come to see implications for this project that go far beyond math class. The center of this democratic mathematics project is the place where thinking about the nature of mathematics, its teaching and learning, and the wider purposes of schooling meet. Stated concisely, our project is about what happens when the math classroom is taken seriously as a place where the civic and democratic aims of schooling are addressed. The potential benefits are considerable for both the learning of mathematics and the health of democracy.[1]

A few words about the state of our democracy are in order here. Starting with a look at democracy narrowly construed, it is very troubling that approximately 100 million Americans (roughly 43% of those eligible) did not vote in the 2016 US presidential election. Perhaps more troubling is the lack of faith in institutions crucial for a functioning democracy that has led to the proliferation of questionable information and questionable sources. Newspapers and other media, governmental agencies, and other important sources of the trusted information that make it possible for democratic participants to make sound decisions have been systematically undermined. It started long before the Trump administration but has certainly intensified since the 2016 presidential campaign season and the four years that followed.

This crisis in confidence has made it possible for Americans to be bitterly divided and for many to turn against the very idea of a public sphere. As we write these words, a global pandemic rages and no one knows how it will alter our futures, though we know it most assuredly will. In the US there is a sharp divide between faith (or at least hopefulness) and skepticism about whether scientific and other forms of expertise can or should inform the policies created to combat the COVID-19 virus.

The politics surrounding the pandemic response in the US offer a particularly interesting way to begin thinking about our perception of mathematics, how we teach and learn it, and our wider social lives. Democracy and expertise have always existed in a somewhat uneasy tension. Early efforts at democracy often involved trying to find ways for common folk to know enough to vote, but little else. For example, Thomas Jefferson's plan to establish three years of public schooling at the taxpayers' expense (for all *free* children, both male and female) was designed to ensure that voters would have the knowledge and faculties to elect wise leaders and to participate in the economy. Jefferson also saw these schools as a means to identify future leaders and his plan included continued free education for the "brightest." He saw this as a way to replace what he called the artificial aristocracy (those who were selected due to their family status and wealth) with a natural aristocracy (those with virtue and talent). That he described his future leader identification system as a way that "twenty of the best geniuses will be raked from the rubbish annually" demonstrates that he had a limited idea of what democracy would come to mean in the future. Jefferson is discussed here not as an exemplar of democratic thinking. Instead his proposed education bill is offered as evidence of the limited scope of what democracy entailed, historically. Historian Benjamin Barber's *An Aristocracy of Everyone* (1992) updated and extended Jefferson's vision of democracy to include concern for the dignity and quality of experience and participation for all.

Contemporary understandings of democracy, at least in theory, tend to be wider and more inclusive. John Dewey, often referred to as *the* philosopher of

American democracy, always highlighted the ways in which voting and the machinations of government were just one part of democracy. To Dewey, democracy connoted the way that we choose to live together, the responsibilities that we recognize toward each other, and the common good. Dewey's democracy also features the promise that one "return" on our commitment to the common, public good is that these arrangements will also allow for individual flourishing.

Our current moment is one during which faith in the "public" is frayed. Furthermore, the tension between individual choice and expert/scientific opinion, while always present, has been politicized to the point where, to many, expertise is seen as an impediment to understanding and sensible action. The version of mathematics education that we develop in these pages, while certainly not the solution to these problems, works to relieve rather than exacerbate these tensions. We describe a way of thinking about mathematics that leads to a democratized school math experience for students, one that is social as opposed to individual, that values skills of reasoning and judgment rather than just rote learning. It is a way of thinking that develops students' capacity and eagerness to use math throughout their lives—solving problems, accomplishing goals, questioning and investigating phenomena.

The philosophy and related practice of school mathematics that we develop and advocate for in these pages, Democratic Mathematics Education (DME), is intended to serve as an ingress, not an obstruction, to the wider world. As we envision it, DME seeks to bring about the development of a school mathematics that is not just open to all, but potentially of interest to all, where people see mathematics as potentially *theirs* and as something that matters in their lives and communities.

We see this moment as providing some great opportunities for democratic education—at the very least there is a great need for it and, in some circles, people are beginning to realize that. The heart of our project involves thinking about how math class can be a site for democratic education and, even more, how this democratic turn can enliven young people's school math experience, helping them to develop critical faculties and agency crucial to robust civic and democratic participation and the skills and knowledge to meaningfully act on their world.

While taking math class seriously as a space for democratic education is the main focus of this book, there are other benefits to the approach we develop. We are also talking about democratic education in the sense that it is education for everyone, for the people. Labaree (1997) convincingly develops democratic equality as a major purpose of public school. To Labaree, democratic equality means much more than just civic participation; he has his sights set on opportunity and equity. Labaree's "democratic equality" tightly intertwines participation and opportunity under the banner of democracy. Opportunity, as it currently operates, is in large part a function of success in mathematics. Unfortunately, today's math class is often used as a sorting mechanism, a decider in the game of who gets to be—by virtue of success in the insular school math competition—thought of as smart and thus given scarce educational and professional opportunities.

This project therefore seeks to stake a claim to mathematics education for everyone,[2] that is, it is an attempt to think in a new way about mathematics in order to democratize math and math education. We draw heavily on Deweyan progressive/democratic education, in both methods and purpose, to develop the pragmatic/democratic philosophy of mathematics education at the core of this project. This rethinking of mathematics can support the needs of critical and social justice-oriented mathematics educators. We develop a way of thinking about mathematics that aligns with student interests, with issues of the day, that is more engaging and inclusive than the typical ways of thinking about school mathematics. In short, the way we conceptualize mathematics in the following pages has the potential to be a democratizing force in a field that is all too often overly abstract, hierarchical, and exclusionary. This pragmatic/democratic philosophy of mathematics invites broad participation and provides a philosophical foundation for a school math experience that can foster sustained engagement from those who have traditionally been discouraged from striving for success and even possessing interest in school math.

The sorting function of school mathematics happens through a process of continual attrition. Those who succeed and who somehow happen to see the connections between school math and their lives are rewarded: the few, the interested, the successful in math.[3] Instead, our hope is that remaking math in a democratic spirit will be democratic not just in terms of teaching the skills of democracy, it will also be democratic in its enactment and in its attention to accessibility, opportunity, and equity. This is a different use of the term democracy, and here we are invoking the version that touts an *opening up* of a field or area. For example, there has been much ink spilled about the power of the Internet to democratize various enterprises. The democratization of the music industry has, the thinking goes, occurred because—thanks to the Internet—musicians can now sell their music directly to the public, sidestepping the monolithic, capitalist beast called the record company. For now we will leave the problems with this form of democratization untouched other than to say that if the new democratized music model makes it technically possible but highly unlikely that an unknown artist's music will be heard far and wide, then this is not the kind of democracy we should be looking for. Democratization is marked by a broadening, not a narrowing, of opportunities, a seeking of all that life can offer. Similarly, democratic mathematics education has the potential to open itself to everyone as well as the potential to open everyone to new and varied possibilities.

It should be fairly obvious how improving access and equity in mathematics education can benefit society, as getting more students further along in mathematics has promise to lead to increased expertise, innovation, general math literacy, to name just a few. We also want to call attention to how this democratization, this opening up of mathematics, will be good, not just for wider society but also for the field. Rochelle Gutiérrez (2002) says it well: "The assumption is that certain people will gain from having mathematics in their lives, as opposed to the field of mathematics will gain from having these people in its field" (p. 147). She goes on to point out how many researchers in mathematics education operate with a deficit view of those previously prevented from full participation in mathematics. Instead, she points out that mathematics is in need of change: "Such programs seem to imply that the people being served by the programs need to improve but that the mathematics does not" (p. 147). In the end, it is not just that the people need mathematics, mathematics needs the people.

Mathematics Education's Untenable Dualisms

Historically, mathematics has been regarded as a bastion of certainty. The characterization of mathematics as objective, logical, and based on objects or truths outside of human involvement has made it resistant to the conceptual shifts that have affected other subject areas. Recent reform efforts that have sought to include social facets of mathematical knowledge within mathematics education have worked against a particularly entrenched understanding of the discipline. Reformers have positioned themselves against these rigid conventions by offering a version of mathematics that is subjective, relative, and fallible. The "math wars" have been raging for several years and show no signs of letting up, pitting traditionalists—those calling for more rigor and a "back-to-basics" approach to mathematics education—against constructivists—those advocating a child-centered, applied approach to mathematics education.

On the surface, there are somewhat similar conflicts occurring in other areas of the curriculum. Certainly, one need not look any further than language arts to see that mathematics is not the only subject area in which a polarized controversy exists. The phonics versus whole language debate, still hotly argued today, seems at first glance analogous to the confrontation at the heart of the math wars.[4] However, there is at least one important difference. While the phonics versus whole language debate is primarily about teaching methods, the math wars—though certainly concerned with teaching methods—is fundamentally a philosophical confrontation. Disputes about how best to teach mathematics are secondary to a more fundamental struggle: traditionalists and constructivists argue over the very nature of the subject matter.

Traditionalists tend to view mathematics as certain, permanent, given, and independent of human activity. Constructivists, on the other hand, focus on the ways in which humans actively create mathematical knowledge, rendering mathematics contingent on human activity. A simple yet powerful way to characterize this split is to borrow from philosopher Richard Rorty's (1999) distinction between those who view phenomena as found versus those who view them as made, told from the point of view of a proponent of the "made" perspective:

One way to describe this impasse is to say that we so-called "relativists" claim that many of the things which common sense thinks are found or discovered are really made or invented. Scientific and moral truths, for example, are described by our opponents as "objective," meaning that they are in some sense out there waiting to be recognized by us human beings. So when our Platonist or Kantian opponents are tired of calling us "relativists" they call us "subjectivists" or "social constructionists." In their picture of the situation, we are claiming to have discovered that something which was supposed to come from outside us really comes from inside us. They think of us as saying that what was previously thought to be objective has turned out to be merely subjective.

(p. xvii)

Rorty's characterization of the found–made distinction in *Philosophy and Social Hope* may, at first blush, appear to be purely academic in nature, but this is not so. As the title suggests, Rorty has an ambitious social agenda for his work, as he views this philosophical stalemate as a hindrance to human progress: "These dualisms dominate the history of Western philosophy and ... the vocabulary which centres around these traditional distinctions has become an obstacle to our social hopes" (p. xii).

Rorty contends that the found–made dualism has constrained our ways of thinking and acting in areas beyond academic philosophy. Dualistic thinking refers to the tendency to conceive of phenomena as splitting into exactly two opposing realms. Often, it is quite useful to draw such distinctions, but at times those once useful lines can become harmful. This is Rorty's complaint about the found–made distinction. He argues that our ability to think is unduly bounded by having to think in such strict either–or terms. Furthermore, over time, entrenched dualisms can make it difficult to understand or even recognize viable alternatives. So habitual is this kind of rigid, dualistic thinking that it may be tempting to categorize our proposed alternative vision of mathematics into one of two existing options. But we ask for a different way of thinking, for resistance to such facile categorization, in an effort to point mathematics education in a different direction.

Extending this idea further within mathematics education, the found–made and other related dualisms create conflicts and prevent the development of new ways of thinking. As a result, this project is organized in such a way as to accentuate and show the damage wrought by the historical splits in the disciplines that affect mathematics education. While exploring the philosophical underpinnings of each side of the math wars we emphasize how each camp suffers from its own sets of polarizations, such as internal–external and individual–social divisions.

Other dualisms are also at work. For example, tensions between belief and knowledge are manifested in different ways throughout this project. Some theorists underemphasize the importance of individual or subjective belief to such an extent that it is virtually ignored, while others argue that objective knowledge is an impossibility. Still others view knowledge and belief differently, focusing on the private and idiosyncratic nature of belief and the public character of knowledge.

With this project, we develop an alternative conception of mathematics for educators and students, a way out of the dualistic trap of being forced to understand mathematics as strictly made or found. Likewise, exposing the impact of social goals and agendas is a critical component of our work, as such exposure helps demonstrate that the lines of the math wars often have been drawn by self-interested parties and did not necessarily arise because they are inherently related to the content of the discipline or the needs of those it serves. Converting hardcore supporters of either side of the math wars is not the primary aim of this project. Instead, the more modest hope is to soften the entrenched sense of "either/or-ism" endemic to conversations within philosophy and mathematics education and to suggest another way of thinking, a way of thinking that can lead to educational and social betterment.

It is a deeply embedded assumption that mathematics consists of found items and that the primary mission of mathematics education is to help children come to understand and learn how to work with them. As noted earlier, most if not all other subject areas have benefited from the realization that social forces and historical contingency play a role in determining their subject matter and that the very subject matter itself is, to varying degrees, socially constructed. With mathematics education, it seems that the traditional belief that the subject exists independently of human activity has been particularly intractable and that as a result, the competing social constructivism within mathematics education is a particularly stubborn and overbearing strain. In a climate of confrontation and hyperpolarization, it is far more difficult for thoughtful and beneficial reform to occur. The first step in moving past this destructive standoff is in rethinking the very nature of mathematics.

Reconceptualizing Mathematics

Each side of the math wars offers valuable insights, but each side is incomplete. An aim of this project is to establish that it can be useful to think of mathematics as partly made and partly found and that the most critical component of the discipline is in the changing patterns of *interaction* between what is given and those who construct mathematical understandings. In other words, it is the mathematical *activity* of those endeavoring to solve genuine problems by engaging with the physical and social worlds that has traditionally been left out of what is considered mathematics. This activity, we argue, should be recognized as a legitimate facet of mathematics.

In order to establish this notion, first we need to detail the predominant underlying philosophical positions that support both the traditionalist and the constructivist ways of thinking. The alternative perspective that we articulate in this project, while less prevalent for philosophy of mathematics, has assumed various forms in general philosophy. This project does not so much set out to forge a brand new way of thinking as much as it borrows from several subgenres of philosophy, stitches them together, and finally introduces them to the ongoing conversations within philosophy of mathematics and mathematics education.

This project is a hybrid. Our ultimate aims are to affect the way that mathematics teachers and students think about—and hence, do—mathematics and mathematics education. In order to affect this change in thought however, this project will necessarily dwell within the realm of philosophy, particularly the philosophy of mathematics. The waters get muddied a bit when considering the emerging field of philosophy of mathematics education. One could see this project as residing almost solely within the parameters of philosophy of mathematics education, since the nature of both mathematics and mathematics education is of central import. It would be a mistake, however, to allow this inquiry to be limited by such a label. Cross-disciplinary contributions are at the heart of philosophical projects of this nature. In *Foresight and Understanding*, Stephen Toulmin (1961) contrasts the perspectives of the disciplinary expert and the philosopher by referring to the expert's "insight" and the philosopher's "outsight." He describes philosophical "outsight" as a way to think about an enterprise from the position of the interested outsider. *Foresight* is an attempt to "focus on science something of the insider's judgment and the outsider's breadth of vision alike," claiming metaphorically that "there is only one way of seeing one's own spectacles clearly; that is, to take them off. It is impossible to focus both on them and through them at the same time" (p. 101). We start with the assumption that from the "outside" it is easier to see that broadening our understanding of what is to be considered mathematics will be beneficial for mathematics education. The outsight that Toulmin speaks of requires that we refuse to settle too comfortably within one discipline and, in addition to a general philosophical outlook, this project will borrow liberally from other disciplines, such as anthropology, psychology, sociology, and history, to name just a few.

A Few Notes on How to Read This Book

Readers deserve a good reason for why they should read any book, and ours is no exception. In order to make the case for DME, it is necessary first to understand the ways of thinking about mathematics that have undergirded mathematics education in the past. A historical account of the roots of the math wars, presented in Chapter 2, provides a context for the contemporary debate. Following this review, in Chapters 3 and 4 we describe the two predominant ways of conceiving of mathematics—absolutism and constructivism—arguing that these two philosophical conceptions undergird the ways of thinking prevalent in each camp of the math wars. Our treatment of absolutist and non-absolutist perspectives is rigorous and includes both the strengths and weaknesses of each philosophy. Chapter 5 is the longest chapter and, one could argue, the one most immediately removed from the teaching and learning of mathematics, but it is also a crucial piece of our development of DME. We provide sustained engagement with relevant facets of evolutionary theory and link them to Dewey's philosophy of mathematics education, while establishing that evolutionary theory has a place in philosophy of mathematics and mathematics education. As we make clear in the first part of the book, mathematics has been very resistant to influence from evolutionary thinkers. As such, we cover a lot of ground in our attempt to make up for a lack of dynamic evolutionary thinking in the philosophy of mathematics, from the Ancient Greeks onward, and for mathematics having been largely ignored by the Darwinian revolution since the late 19th century. That other domains have had that much time to integrate Darwin's contributions should make clear the ambition (and length) of this chapter. Once familiar with evolutionary thinking, readers are primed to engage with DME. The later parts of the book turn more toward how the various ways of thinking about mathematics play out in the classroom and how DME fits into the contemporary mathematics education scene. Chapter 6 provides a direct explication of DME. In Chapter 7 we consider some potential educational implications of the adoption of the three approaches. The recent critical mathematics education movement is then presented in Chapter 8, and the relationship between it and DME is explored. The book concludes in Chapter 9 with discussions about empowerment in mathematics education, the worth of engaging in humanities-oriented projects related to school and school mathematics and how they help us to recognize the connections between math and morality/values, and some conclusions about the hope and promise of the democratic mathematics classroom.

People read books for a variety of reasons. We wrote this one under the assumption that most readers will be as excited about the philosophy of mathematics education as we are. Furthermore we see the book as an evolving argument, with each chapter supporting the next, and we believe that the credibility and impact of our overall argument is fortified by the chapters in which we lay the historical and ideological groundwork. While our book is hardly the impenetrable tome that is Russell and Whitehead's *Principia Mathematica*, we humbly acknowledge that Chapters 3, 4, and 5 are dense and difficult. But just as we endorse the agency of students, we call on the reader to read with intention and strategy. We invite them to survey the textual landscape of each chapter in advance, to take advantage of the diagrams and tables provided as a means of mentally organizing the many theories and theorists, and perhaps even to thumb through the headings in advance to anticipate where interest may be greatest. Again, we believe that reading about the different ways math has been perceived is ultimately a fruitful endeavor, and the introduction to democratic mathematics education, when it is at last presented, will be all the richer for having done so.

Notes

1   We are based in the US and while this brief discussion addresses the American context, many of the same problems are endemic to democracies around the world, from the UK's Brexit debacle to the rise of autocracy and fascism in Eastern Europe, to name just two of the disturbing trends.
2   We are sensitive to the critique that phrases such as "mathematics for everyone" can overlook issues of race and racism and may be interpreted as assimilationist (see Martin, 2009, p. 303). Bearing this in mind, we turn our attention to developing a robust democratic approach that we believe can, by virtue of its philosophical underpinnings, invite wider participation not just in its consumption but in its continual recreation. We thus envision a mathematics *for* everyone and *of* everyone.
3   Success in mathematics need not be restricted to the world of applied math. There is another path to success, namely, embracing the mathematician's delight in the abstract beauty of theoretical math.
4   The "reading wars" have a surprisingly long history. They began over 200 years ago when Horace Mann first condemned the practice of teaching letter–sound associations (Castles et al., 2018).

References

Barber, B. (1992). *An aristocracy of everyone: The politics of education and the future of America.* New York: Ballantine Books.

Castles, A., Rastle, K., & Nation, K. (2018). Ending the reading wars: Reading acquisition from novice to expert. *Psychological Science in the Public Interest*, 19(1), 5–51.

Gutiérrez, R. (2002). Enabling the practice of mathematics teachers in context: Toward a new equity research agenda. *Mathematical Thinking and Learning*, 4(2–3), 145–187.

Labaree, D. (1997). Private goods: The American struggle over educational goals. *American Educational Research Journal*, 34(1), 39–81.

Martin, D. (2009). Researching race in mathematics education. *Teachers College Record* 111(2), 295–338.

Rorty, R. (1999). *Philosophy and social hope.* New York: Penguin Books.

Su, F. (2020). *Mathematics for human flourishing.* New Haven, CT: Yale University Press.

Toulmin, S. (1961). *Foresight and understanding: An enquiry into the aims of science.* New York: Harper Torchbooks.

2

MATHEMATICS EDUCATION IN HISTORICAL AND CONTEMPORARY CONTEXTS

"First, we were supposed to say 'carry the one,' and now we're *not* supposed to say 'carry the one.' It keeps changing." This grievance, aired by an exasperated parent immediately after learning that one of the authors was a school mathematics specialist, may sound familiar and uniquely modern. After all, many parents today might relate to her frustration with her child's math curriculum. The perceived shift from procedural to conceptual learning seemed to happen overnight. What was once called "one" suddenly became one ten, one hundred, or one unit of some power of ten[1] in an effort to infuse meaning into the meaning-less algorithms that dominated elementary mathematics for generations. But the struggle between traditional and reform mathematics is, in fact, a contemporary expression of an age-old structure, one with a story worth telling. We begin our historical survey at the turn of the 20th century, an era marked by industrialization, progressivism, and a new form of psychology that would challenge and eventually dismantle the pseudoscientific constructs of an earlier age.

Roots of the Math Wars

In the early 1900s, some mathematics educators were beginning to question the soundness of faculty psychology and its requisite "mental discipline" justification of mathematics education (Kliebard, 1995, p. 92). Faculty psychology, the notion that general mental competencies or "faculties" could be developed by work in certain subject areas "was based on two metaphors that were useful in planning curriculum (mind as muscle to be exercised and mind as vessel to be filled)—(it) emphasized mental 'faculties' such as memory and attentiveness" (Tozer et al., 2002, p. 38). Faculty psychology's roots run deep, having been traced back to the Ancient Greek idea that studying geometry led to increased general intelligence

(Kliebard, 1995, p. 4). The mental discipline rationale for mathematics education held that rigorous work in memorization and deduction of mathematical facts and rules led to sharper, clearer thinking in general. As such, mental disciplinarians placed a great deal of faith in the ability of individuals to transfer what they learned from one task to another. This far-reaching notion of transfer had been under fire since the late 1800s, beginning with research that cast doubt upon the existence of a discrete faculty called memory.[2] In an effort to better explain the process of transfer, E.L. Thorndike formulated and popularized the psychology of associationism. Rather than emphasizing general capacities (such as memory and reasoning), Thorndike posited a mind whereby individual connections or associations were made between discrete bits of information. In *The Psychology of Arithmetic* (1924), as well as other works, Thorndike applied his associationist psychology directly to mathematics education. Chapter V bears the telling title, "The Psychology of Drill in Arithmetic: The Strength of Bonds." In it, Thorndike argues that students fail to perform well in mathematics primarily because they have not formed the proper mental bonds: "The constituent bonds involved in the fundamental operation with numbers need to be much stronger than they are right now. Inaccuracy in these operations means weakness of the constituent bonds" (p. 102).

Associationism was offered as a replacement for faculty psychology: "Thorndike concluded that school subjects are to be valued for their specific content and not for any generalized disciplinary powers" (Karier, 1986, p. 171). Interestingly, the mathematics classes built upon associationism were not much different from the classes that preceded them, as rote memorization of mathematics facts was the primary pedagogy advocated by Thorndike.

The Progressive Education Association (PEA) was an organization made up of educational scholars and practitioners committed to moving schooling beyond traditional approaches, often advocating for modern child-centered education. Their stance on mathematics education was motivated, at least in part, by a desire to counteract the classroom practices suggested by Thorndike (Ellis & Berry, 2001, p. 3). Student interest and experience became the focal point of the PEA-influenced mathematics classroom, as opposed to the emphasis on individual mathematical facts and student habits (English & Halford, 1995). The PEA's work is significant to our project, as it marks the origins of a pattern that led to the development of two sharply divided points of view with respect to mathematics education embedded in the math wars. The pattern is one whereby groups of stakeholders come to identify themselves by what they are reacting *against* as much as by any constructive ideas related to what they stand *for*.

The "new math" of the 1960s was the next major development in mathematics education. Contrary to common conception, the new math was not a single program, but rather a gradually developed way of thinking about mathematics and mathematics education (Klein, 2001, p. 6). It reflected the growing influence of professional mathematicians who often emphasized the abstract and logical facets of mathematics, in contrast to the progressive educators' tendency to place the students at the center of mathematical activity.[3] Back in 1950, the Harvard Committee's *General Education in a Free Society* defined mathematics in a way that highlighted the influence of professional mathematicians and their focus on branches of mathematics that were not readily applied to our everyday lives: "Mathematics may be defined as the science of abstract form" (Harvard University, 1950, p. 161). This definition helped to usher in the new math movement.

When, in 1957, the Soviet Union took the lead in the space race by launching *Sputnik*, the growing sentiment that American schools had become "soft" in the wake of progressive, child-centered reforms merged with a newfound desire to use the schools as a way to compete with what was perceived as Soviet mathematics/science superiority. The emphasis of the new math on abstract, theoretical mathematics was the nation's solution to the question of how to return rigor to the mathematics curriculum. The ways in which the new math focused on a set curriculum and tended to emphasize the disciplinary style of expert mathematicians was also in line with the wider political climate of the time. The relative conservatism of the Eisenhower era coupled with an increasing faith in the importance of expert knowledge[4] intertwined to produce a congenial environment for the development of the new math movement.

The excitement for new math reforms fairly quickly gave way to the realization that many students and parents ended up confused and left behind by the movement's tendency to push advanced, abstract, and less overtly useful material deep into the K-12 curriculum.[5] The broad social changes in America in the 1960s most likely also contributed to the failure of the new math to catch on and endure in the nation's classrooms. By the late 1960s, the general suspicion of authority that had started in the nation's universities had worked its way down into the high schools and middle schools in the US.

The new math's reliance on and reverence of the expertise of professional mathematicians (as opposed to a focus on everyday mathematics) did not fit well with the wider anti-authority movement. Furthermore, the unrest of the late 1960s and early 1970s led to an increase in elective courses for university and high school students: "The intense public criticism of schools in recent years, together with widespread student unrest during the late 1960s, gave rise to a demand for 'relevant' educational programs that would more closely approximate students' needs" (Bredo & Bredo, 1975, p. 450). Consequently, courses of study in mathematics were often interrupted, as there was a perception that the rigid structure of mathematics course sequences was an unwelcome imposition from authority. These interruptions made it difficult for new math programs to continue.

In the years since the initial backlash to the new math movement, some forces within mathematics education have sparked a renewed interest in the progressive tenet of considering the learner's involvement in the learning process, as opposed to the more narrow focus on the content of mathematics. The National Council of Teachers of Mathematics (NCTM) has worked to develop a mathematics education that is grounded in the belief that student involvement within mathematics-rich contexts is crucial to the successful teaching and learning of the subject.[6] Interestingly, while the threat of the new math is gone, the movement (led by the NCTM) that grew out of a reaction to it is still reacting against other mathematics education interests. One explanation of the origins of today's traditionalist or "back-to-basics" camp recognizes the group's formation as largely a response to a perception that the NCTM and other like-minded groups have helped the pendulum swing too far away from mathematical rigor.

Today's Context: Inadequacies in the Pendulum Metaphor

The origins of each group in today's math wars are not universally agreed upon. For example, Ellis and Berry conceive of the traditionalist movement as originating in reaction to perceived shortcomings of the new math reforms of the 1950s and 1960s (2001, p. 4). This differs from the account just presented (and backed by Klein and the *Mathematically Correct* website) which situates the origins of the back-to-basics group within a reaction to more recent NCTM-backed reform initiatives, as well as to the general social changes of the 1960s. If one interprets this very brief history's theme in a logical, but equally dualistic way—that mathematics curriculum debates historically have followed the pattern of a metaphorical pendulum swinging back and forth from traditional notions of mathematics emphasizing content to different, usually child-centered ideas of how best to teach mathematics—then the roots of both groups in the contemporary debate are so deep that it is difficult, if not impossible, to trace their origins. In other words, determining how and when the back-to-basics and reform groups got started may have more to do with how, when, and why the researcher has jumped into the debate than with any objective version of what actually happened.

Even this brief history of the math wars reveals a pattern that is far too complex to be adequately explained by the traditional pendulum metaphor. In "The Social Construction of Learning," Eric Bredo provides a history of the development of different American learning theories. According to Bredo, there is a historic tendency toward polarization between overly physical (behaviorism, for example) and overly mental (cognitivism) conceptions of learning. Rather than explain the historic developments of learning theory as one school of thought reacting against another in a pendulum-like manner, Bredo posits that learning theories gain respectability for three reasons: 1. A given theory was a good fit with the prevailing political climate; 2. The theory helped advance professional claims in the field; 3. The approach addressed practical problems that had become salient at the time (Bredo, 1997).

The main point that we are trying to make is *not* that the origins of the two groups are forever lost to the hopelessly subjective nature of historical research;

indeed historical endeavors are important, enlightening, and not inherently overly subjective. Instead, we are claiming that the history of the debates about how best to conceive of the mathematics curriculum—in spite of the decidedly un-pendulum-like complexities suggested by Bredo's analysis of the recent history of learning theories—shows a remarkably consistent tendency to break into two distinct and oppositional camps.[7] As a result of this dichotomous structure, those wishing to improve mathematics education typically end up either thinking along the well-worn lines of one of the camps or having their ideas placed there by others. Consequently, the passing of time has obscured much of the context of the debates and rendered most historical accounts of mathematics education flat and inaccurate. Looking back we see a pendular motion to the changes in the field when really a more dynamic story is being compressed and obscured. Complex social phenomena (from changing social, geopolitical, and cultural winds to breakthroughs in science, mathematics, psychology, philosophy, education, and other disciplines) have affected the aims and methods of school mathematics. It seems that the dualistic found vs. made type of thinking has fostered an environment in which the only avenue to expression is through consigning ideas and actions to one side or the other. The result is that our ability to think about mathematics and mathematics education is constrained by this "two-option" mentality.

There have been noteworthy attempts to counter the sharp polarization within mathematics education. In "Conceptual and Procedural Knowledge in Mathematics: An Introductory Analysis" (1986), Hiebert and Lefevre, after a careful explanation of the nature of what they refer to as "conceptual" and "procedural" knowledge, work to show how the two are not necessarily opposed and that they actually depend on each other to a certain degree. Briefly summarized, according to Hiebert and Lefevre, procedural knowledge involves the rules and skills of mathematics and conceptual knowledge refers to forming links between new and existing knowledge. The authors acknowledge their procedural–conceptual distinction is related to rote versus meaningful learning and Scheffler's (1965) split between "the propositional use of 'knowing that' and the procedural use of 'knowing how to'" (Hiebert & Lefevre, p. 1). In a sense, this work seeks to combine the traditionalist preoccupation with learning a set body of rules, skills, or procedures with the reform-oriented push toward understanding.

Likewise, the 1980s focus on problem-solving as a main mission of mathematics education sought to make procedures and concepts work together in an effort to unify mathematics education subject matter. In "Learning to Think Mathematically: Problem Solving, Metacognition, and Sense Making in Mathematics" (1992), Alan Schoenfeld details the contemporary emergence of problem-solving from its roots in George Pólya's *How to Solve It* (1957), to the NCTM's *Agenda for Action* (1980), to the prominent place afforded it by the NCTM today. Unfortunately, the math wars' polarizing tendencies have made it difficult for a middle ground to be forged and problem-solving approaches to mathematics education tend to be simply funneled into the reform mathematics camp and not typically seen as an alternative to the dualistic thinking so prevalent in the field.

One reason why attempts to blur the damaging distinctions within mathematics education—such as the work of Hiebert and Lefevre, Schoenfeld, and others—have not led to wider application is that there is a more fundamental polarization present.[8] That is, work seeking to show the connectedness or continuity between concept and procedure is taking place entirely within the realm of mathematics education. We argue that while concept and procedure should be conceived of in a more integrated way, there are untenable dualisms operating on a philosophical level that affect what transpires within mathematics education.[9]

Stephen Lerman, drawing on Basil Bernstein's concept of recontextualization, suggests that there are three levels of knowledge in mathematics education:

> At the first level are the surrounding (sometimes called foundation) disciplines of psychology, sociology, philosophy, anthropology, (in our case) mathematics, and perhaps others. At the second level are mathematics education and other curriculum areas of educational research. At the third level are curriculum and classroom practice.
>
> *(Lerman, 2000, p. 20)*

Lerman goes on to explain that recontextualization occurs when ideas from one level are adapted to and used on another level. While we are not sure whether differing levels is the best way to conceive of these distinctions—perhaps they could be conceptualized as overlapping domains—Lerman is quite right to recognize that there are differing ways of thinking about and dealing with mathematics that may or may not move between levels or domains. Our idea is that a "foundational" split exists philosophically between those who see mathematics as absolute, extra-human, and fundamentally found (most traditionalists) and those who think of mathematics as a psycho-socio-cultural entity (most reformers). This project offers an alternative to having to choose between philosophical positions. These Bernsteinian recontextualizations can help ease tensions endemic to the overly dualistic ways of thinking existent in the domains of mathematics teaching and learning.

Toward an Evolutionary Philosophy of Mathematics

The interaction between organisms and their environments is a critical component of evolutionary theory. The ways in which people interact with their environments are numerous and they are both psychological (in the form of an individual in the world) and social (humans are inherently social and most of the ways in which we interact with our environment are related to larger groups). The point is that there is something external to people and groups with which we interact that helps co-construct our understanding of the world. Mathematics is, we posit, part of our world. It stands to reason that a useful philosophy of mathematics will acknowledge both the internal and external elements of mathematics and, perhaps most importantly, emphasize the interaction between them. To go one step further, what each element is partially depends on the other, as the interaction to which we refer leads to fundamental changes in both. In essence, the external and internal elements have a hand in co-creating each other. In doing so, the made–found and internal–external splits are rendered less distinct.

Forging such a "useful philosophy" is no small or simple undertaking, and our choice to engage with evolutionary theory to get past the troubles that come along with existing ways of thinking about mathematics may strike the reader as relatively novel[10] or even strange. We do so for several reasons. Interactions between individuals, groups, and environments create different mathematical constructions; some seem to work better than others. The more fruitful help us live our lives better and also sometimes lead to extensions of mathematical thought. In this sense, our conception of mathematics has much in common with the Darwinian understanding of the formation of species by natural selection. Furthermore, environmental and cultural contexts constrain the possibilities of mathematics similarly to the ways in which they constrain biological evolution. In other words, both the absolutist's mathematical entities (whatever they may be) and the contingencies created by our human existence shape the directions that mathematics may take and strongly affect the field of mathematics.

Additionally, philosophies of mathematics in general can be thought of as the product of an evolving discussion. The traditional camp's undergirding absolutism has roots in ways of thinking that clearly predate evolutionary theory in that its version of mathematics is fixed, static, and essentialized. The heart of the reform camp's philosophical position employs some post-Darwinian thinking (the Wittgensteinian rejection of essentialism and the emphasis on the history of change in mathematics, to cite a pair of examples). Conceiving of the two camps as philosophies that were useful given their particular contexts, it does not seem too much of a stretch to imagine a need for the development of new and different conceptualizations that suit our contemporary context. To continue to draw on this Darwinian metaphor, this new way of thinking will not be a spontaneous creation, but will instead be an updated philosophy that retains strengths of the old while adapting to the current environment.

As mentioned in Chapter 1, it is our hope that this philosophical consideration of mathematics and mathematics education can help to show that the entrenched ways of thinking about (and actually doing) school mathematics are in need of change. Furthermore, this is not merely a methodological dispute. Throughout these first two introductory chapters we have described the underlying tension in the math wars as philosophical. While we think this is true, there is also a moral component to this situation. This dispute is also about differing aims and values as well as differing epistemologies or ontologies. We hope that our philosophical analysis can help sort out and clarify the claims of the involved parties. An analysis of not just what is thought, but *why* it is thought (normatively speaking), will help us to consider what current goals for mathematics education are and what they ought to be.[11]

Summary of Purpose and Thoughts on Method: Rediscovering the Bonds between Math and Humanity

Our purpose is to extend civic/democratic education into our nation's math classrooms in order to help with the teaching and learning of mathematics and also to educate in such a way as to increase students' ability to meaningfully participate in the social and political aspects of our democracy. Perhaps even more fundamentally, we see school mathematics as designed and practiced in ways that all too often obscure the organic connections between math and humanity—our values and morality. The ways in which mathematics is frequently conceptualized exacerbate these problems. As a result, reforms (some promising, some not so promising) are justified by radical and difficult to defend philosophical stances. In Part II—*The Nature of Mathematics*, we reconsider the philosophical support for some conceptions of mathematics (both absolutist and fallibilist), keeping what is useful and discarding what is not. In Part III—*The Alternative*, we draw on the work of scholars from a variety of traditions in order to develop an evolutionary philosophy of mathematics and apply it to school mathematics in a way that squares with the broader democratic aims of schooling.

As for our approach, this is primarily a philosophical study and not an empirical one. As a result, our conclusions will—while possessing implications for practice—be largely theoretical. We see the intellectual range of motion accorded by this type of project as beneficial and perhaps even necessary for any attempt to break free of some of the traditions that have overstayed their welcome. In the book's final section, Part IV—*Enactment*, we explore some of the possible educational consequences of adopting particular philosophies of mathematics. It is here that our philosophical work helps readers to experience a variety of concrete possibilities for school math that did not exist prior to our evolutionary/democratic reconceptualization.

In terms of particular methods, we cite primary and secondary sources in an effort to blend historical and philosophical inquiry. It is not our intent to analytically spell out our premises and conclusions. This is in keeping with our belief that we should be leery of anyone who claims to have proven the one *best* way to teach mathematics or to possess a privileged position in terms of knowing which methods or ideas necessarily lead to the most effective mathematics instruction. We believe that we are working toward the highest degree of utility that an inquiry of this nature can hope to accomplish, namely offering a philosophical vantage point that is useful in contending with current states of affairs and that has promise to guide us in productive future directions. As stated in the previous chapter, we are situating our work as much within the mathematics education literature as the philosophy of mathematics, as it is an aim of this project to offer useful ways for mathematics educators, classroom teachers, and students to think about mathematics. For this reason, we will include practical descriptions, explanations, or other vignettes when they will clarify or deepen the main point.

Notes

1 In the case of adding decimals, such as $1.5 + .75$, sometimes a "one" (as in, 10^0) is indeed "carried" to the next place value.

2 William James was an early doubter of the belief that memorizing improved the memory. He confronted the topic in his *Psychology* in 1895.

3 In *Mathematics: The Loss of Certainty*, Morris Kline (1980) explains the professional mathematics community's increasing emphasis on abstract and formally logical conceptions of mathematics as coming about largely to protect the perfection and majesty of mathematics in light of a series of developments that cast doubt on the traditional notion of mathematics as purely objective and infallible.

4 Burton Clark's *Educating the Expert Society* (1962) details this phenomenon, arguing that American society saw an increasing specialization throughout most of the 20th century and a consequent increase in reliance on experts.

5 In 1962, Morris Kline led the charge of mathematicians and other concerned individuals. Jones and Coxford's "Mathematics in the Evolving Schools" details this part of the story and includes excerpts from an open letter Kline wrote about the negative effects of the new math movement. Also see Kline's *Why Johnny Can't Add: The Failure of the New Math* (1973).

6 From *An Agenda for Action* (1980) to the *Curriculum and Evaluation Standards* (1989) and *Principles and Standards for School Mathematics* (2000), the NCTM has produced a series of progressive or humanistic guides for mathematics education.

7 We are not arguing that the two camps are forever static. Indeed, this short history of mathematics education shows that the two sides are perpetually changing in a very gradual manner.

8 We should stress that this work can be helpful and is well-regarded (as is evidenced by the publishing record of the authors in question) but that it has not helped mediate between traditionalists and reformers. Hence, the math wars rage on.

9 We are not arguing for doing away with all dualistic thinking, as it is frequently important to make distinctions given particular contexts. We are calling for a recognition that we have chosen to draw the lines and that there are other (possibly more productive) ways to think of mathematics than as fundamentally made or found.

10 Dewey's thinking was very clearly influenced by Darwin and this influence even found its way into the relatively obscure work on mathematics education that he authored with James McLellan. Our Chapters 5 and 6 include detailed treatments of this Darwin-influenced Deweyan mathematics education as well as other efforts to engage with evolutionary thinking.

11 Some might argue that philosophy and feelings should be kept separate. One premise of this work is that this separation is an impossibility. How we come to know and think about our world is inextricably linked to who we are—hopes, goals, feelings, and all. William James goes as far as to say that "temperament" is an unspoken but primary contributor to a philosopher's outlook and that ignoring this often leads to certain problems in the discipline: "There arises thus a certain insincerity in our philosophic discussion: the potentest of all our premises is never mentioned. I am sure it would contribute to clearness if in these lectures we should break this rule and mention it, and I accordingly feel free to do so" (James, 1909/1975, p. 11).

References

Bredo, A., & Bredo, E. (1975). Effects of environment and structure on the process of innovation. In J.V. Baldridge & T.E. Deal (Eds.), *Managing change in educational organizations: Sociological perspectives, strategies, and case studies.* Berkeley, CA: McCutchan.

Bredo, E. (1997). The social construction of learning. In G. Phye (Ed.), *Handbook of academic learning: Construction of knowledge* (pp. 3–46). New York: Academic Press.

Clark, B. (1962). *Educating the expert society.* San Francisco: Chandler Publishing Company.

Ellis, M., & Berry, R. (2001). *The paradigm shift in mathematics education: Combining cognition and culture to bring equity and meaning into the equation.* Unpublished paper presented at AESA 2001, Miami.

English, L., & Halford, G. (1995). *Mathematics education: Models and processes.* Mahwah, NJ: Lawrence Erlbaum Associates.

Harvard University (1950). *General education in a free society.* Cambridge, MA: Harvard University Press.

Hiebert, J., & Lefevre, P. (1986). Conceptual and procedural knowledge in mathematics: An introductory analysis. In J. Hiebert (Ed.), *Conceptual and procedural knowledge: The case for mathematics* (pp. 1–27). Hillsdale, NJ: Lawrence Erlbaum Associates.

James, W. (1895). *Psychology: The briefer course.* Notre Dame, IN: University of Notre Dame Press.

James, W. (1909/1975). *Pragmatism and the meaning of truth.* Cambridge, MA: Harvard University Press.

Jones, P., & Coxford, A. (1970). Mathematics in the evolving schools. In A. Coxford, H. Fawcett, P. Jones, H. Karnes, L. Nelson, A. Osborne, & J. Weaver (Eds.), *A history of mathematics education in the United States and Canada* (pp. 11–92). Washington, D.C.: National Council of Teachers of Mathematics.

Karier, C. (1986). *The individual, society and education: A history of American educational ideas.* Champaign: University of Illinois Press.

Klein, D. (2001). A brief history of American k-12 mathematics education in the 20th century. www.csun.edu/~vcmth00m/AHistory.html

Kliebard, H. (1995). *The struggle for the American curriculum: 1893–1958.* New York: Routledge.

Kline, M. (1973). *Why Johnny can't add: The failure of the new math.* New York: St. Martin's Press.

Kline, M. (1980). *Mathematics: The loss of certainty.* New York: Oxford University Press.

Lakatos, I. (1976). *Proofs and refutations.* Cambridge: Cambridge University Press.

Lerman, S. (2000). The social turn in mathematics education research. In J. Boaler (Ed.), *Multiple perspectives on mathematics teaching and learning* (pp. 19–44). Westport, CT: Ablex.

National Council of Teachers of Mathematics (1980). *An agenda for action.* Reston, VA: National Council of Teachers of Mathematics.

National Council of Teachers of Mathematics (1989). *Curriculum and evaluation standards.* Reston, VA: National Council of Teachers of Mathematics.

National Council of Teachers of Mathematics (2000). *Principles and standards for school mathematics.* Reston, VA: National Council of Teachers of Mathematics.

Pólya, G. (1957). *How to solve it.* Princeton, NJ: Princeton University Press.

Scheffler, I. (1965). *Conditions of knowledge: An introduction to epistemology and education.* Chicago: University of Chicago Press.

Schoenfeld, A. (1992). Learning to think mathematically: Problem solving, metacognition, and sense making in mathematics. In D.A. Grouws (Ed.), *Handbook of research on mathematics teaching and learning: A project of the National Council of Teachers of Mathematics* (pp. 334–370). New York: Macmillan.

Thorndike, E. (1924). *The psychology of arithmetic*. New York: Macmillan.

Toulmin, S. (1961). *Foresight and understanding: An enquiry into the aims of science*. New York: Harper Torchbooks.

Toulmin, S. (1972). *Human understanding: The collective use and evolution of concepts*. Princeton, NJ: Princeton University Press.

Tozer, S., Violas, P., & Senese, G. (2002). *School and society: Historical and contemporary perspectives*. New York: McGraw Hill.

PART II

The Nature of Mathematics: Two Competing Perspectives

3

ABSOLUTIST MATHEMATICS

The Infallibilist/Apriorist Thesis

> The world of mathematics … is really a beautiful world; it has nothing to do with life and death and human sordidness, but is eternal, cold, and passionless. To me, pure mathematics is one of the highest forms of art; it has a sublimity quite special to itself, and an immense dignity derived from the fact that its world is exempt from change and time … The only difficulty is that none but mathematicians can enter this enchanted region … And mathematics is the only thing we know of that is capable of perfection; in thinking about it we become Gods. This alone is enough to put it on a pinnacle above all other studies.
>
> *(Bertrand Russell,* The Selected Letters of Bertrand Russell, Volume 1: The Private Years 1884–1914*)*

What a heavy responsibility mathematics seems to have shouldered. Hailed as a discipline of definite and irrefutable knowledge, mathematics has served as a "lighthouse of absolutism" for much of its long history (Ravn & Skovsmose, 2019). While there were and are philosophical challenges to an absolutist view of mathematics—challenges we will treat with equal scrutiny—absolutism has demonstrated remarkable staying power in the protracted debate about the nature of mathematical knowledge.

Absolutism is a family of related approaches constituting the predominant philosophy of mathematics. Also called infallibilism or apriorism, absolutism affords mathematics an elevated status that is distinct from empirical science. Absolutists view mathematics as a unique branch of knowledge that offers certain Truth. Here, the imposing capital "T" reflects not just certainty but an implied eternality. Nothing less than a capital "T" Truth could be expected of the exclusive "enchanted region" Bertrand Russell described in his 1901 letter to Helen Thomas, quoted above (Russell, 2002, p. 224). His paean to mathematics is very much in the absolutist tradition: it conveys an unwavering trust in the stability and certainty of mathematical knowledge.

To the absolutist, mathematics is therefore the "paradigm of infallibly secure knowledge," as Paul Ernest has said (1998, p. 1). This feature, this infallibility, can be traced to ancient origins. It is impossible to overstate the influence of Euclid's *Elements* on the absolutist perspective of mathematics. Written around 300 BCE, Euclid's masterpiece is more than the ur-geometry text many of us imagine it to be. It is an enactment of Aristotelian principles for generating scientific truths. Specifically, it is an enactment of a deductive method—of systematically deriving truths from Truths[1]—that when executed correctly guarantees the irrefutability for which mathematics would long be celebrated. It was the *Elements* that "established (mathematics) as a certain foundation for human knowledge, and other sciences had a standard to measure themselves against" (Ravn & Skovsmose, 2019). So durable was this perspective that nearly 2,000 years later the mathematical work of luminaries such as Descartes,[2] Newton, and Spinoza could be fairly characterized as absolutist (Ernest, 1998). The absolutist view of mathematics found perhaps its most profuse expression in the early 20th century in Russell and Whitehead's *Principia Mathematica* (1910–1913). In three imposing volumes, their development of logicism was intended to "provide a rigorous and certain foundation for all mathematical knowledge" (Ernest, 1998, p. 1). The belief that mathematics could be so distilled suggests a confidence in its supreme structural and methodological soundness. To the absolutist, mathematics is about knowing, where knowing connotes a clear line between that which is certain and that which is not.

The special status of mathematics afforded by absolutism has rarely been challenged (Hersh, 1997; Kitcher, 1983). As Kitcher points out: "Virtually every philosopher who has discussed mathematics has claimed that our knowledge of mathematical truths is different in kind from our knowledge of the propositions of natural science" (1983, p. 3). To ordinary people and philosophers alike, he contends, mathematics is the pinnacle of human knowledge: it is used to judge other knowledge claims, and mathematical knowledge is not obtained by ordinary perceptual experience. This line of thought, which Kitcher labels apriorism,[3] echoes the general sentiments of the typical mathematical absolutist and has its roots in Platonism.

Platonism

In a biography of mathematician Julia Robinson there is an anecdote about her childhood fascination with natural numbers. Robinson recalls, "One of my earliest memories is of arranging pebbles in the shadow of a saguaro, squinting because the sun was so bright" (Reid, 1996, p. 3). We can imagine a young Robinson in the 1920s Arizona landscape, fiddling with arrays of varying shapes and dimensions. What might she have inferred from an arrangement of pebbles? We, too, can engage in some number play here. In so doing, we might notice that square numbers can also be expressed as the sum of two consecutive triangular numbers:

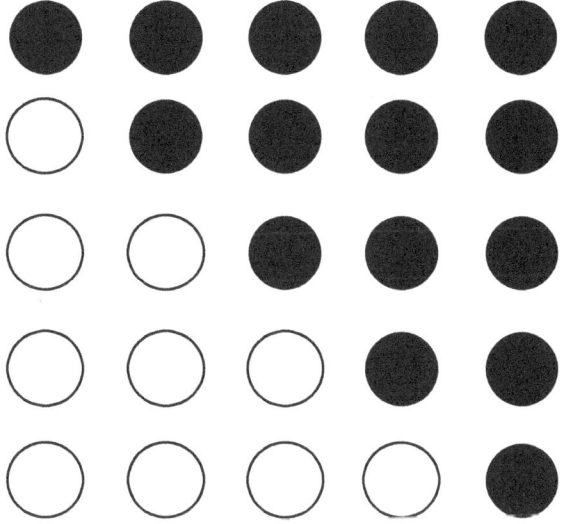

FIGURE 3.1 An Arrangement of Pebbles

The 5 × 5 square array represented in Figure 3.1 can be partitioned into two triangular arrays, as we have illustrated above. The triangular array formed by the 10 white circles represents the sum of the first four consecutive natural numbers (1+2+3+4). Here, we can see why the number 10 is also known as a "triangular number." If we were to add another natural number (1+2+3+4+5), the sum would be the next triangular number, 15, represented by the black circles. The sum of 10 and 15, 25, is indeed a square number. With a list of the first triangular numbers in hand (1, 3, 6, 10, 15, 21, 28, 36 …) we can add any two adjacent members of the list and confirm that their sum is in fact a square number. The relationship among these figurate numbers is indeed generalizable and connected to other mathematical topics such as quadratic functions and combinatorics. But where exactly do these mathematical realities reside? In the array? In us? Or do they exist somewhere else? According to the most prevalent form of absolutism, there was and is a clear answer. Robinson's own words make this point: "We can conceive of a chemistry that is different from ours, or a biology, but we cannot conceive of a different mathematics of numbers. What is proved about numbers will be a fact in any universe" (Reid, p. 3).

The most widespread form of mathematical absolutism, and probably the most adhered to philosophy of mathematics in general, is Platonism (Hersh, 1997; Kitcher, 1983). For Platonists, mathematical truths, even mathematical objects, are fundamentally extra-human. They exist independently of our successes and failures, independently of our conjectures and proofs, independently of us entirely. Per the Platonist, we merely *uncovered* a pre-existing truth about the relationship between square and triangular numbers, just as Robinson uncovered, or found, characteristics of particular Diophantine equations critical to computability theory. The found–made dualism has reared its head rather obviously here. Platonism is a strong representative of the "found" side of the dualism, and formalism, explored later in this chapter, is an absolutist version of the "made" side. In his well-known essay defending the pursuit of abstruse mathematical topics, *A Mathematician's Apology*, G.H. Hardy's Platonist stance is unmistakable: "I believe that mathematical reality lies outside us, that our function is to discover or observe it, and that the theorems which we prove, and which we describe grandiloquently as our 'creations,' are simply the notes of our observations" (1940/2001, p. 123). This is likely not an uncommon perspective among mathematicians. Parsons (1967), in fact, claims that Platonism is the way that most modern mathematicians think about mathematics. However, Hersh (1997) sees Platonism as the way that most laypeople think about mathematics. We tend to agree with Hersh: a latent Platonism is at work when most people think about the nature of mathematics (which admittedly, is probably not all that often).

Plato and Platonism

Plato is generally credited with originating the dualistic way of thinking that has permeated Western philosophy from the time of Ancient Greece until today (Rorty, 1999, p. 12; Press, 1999). Plato's notion of existence was driven by dualisms. Among the many dyads constituting reality were body–soul, appearance–truth, belief–knowledge, the changing–the unchanging, the senses–the intellect, and others (Press, 1999, p. 44). Of particular relevance to mathematics were Plato's two proposed realms, the imperfect transitory realm of our material existence and the perfect realm of ideas, or what he called the realm of "Forms." These two worlds are ontologically, epistemologically, and even temporally distinct from one another. The perfect realm is a world of eternal, knowable reality, while the imperfect realm is home to what is earthly, concrete, and dynamic— the stuff of everyday experience (Quinton, 1967, p. 348). To help distinguish the realms, we have listed some of their salient characteristics:

TABLE 3.1 Perfect and Imperfect Realms

Perfect realm	Imperfect realm
Abstractions, ideas ("Forms")	Concrete
Eternal	Transitory, ephemeral
Unvarying	Dynamic
Real, knowable	Not wholly real
Knowledge	Belief

Note that although objects classified in the imperfect realm are characterized as "not wholly real," we concur with Quinton that this does not mean they are non-existent or even unimportant. But they are nevertheless distinct from objects in the perfect realm. In a discussion of the mind of the philosopher in *The Republic*, Plato distinguishes between the states of "true being" and "becoming" by noting differences between the two worlds: "Let us suppose that philosophical minds always love knowledge of a sort which shows them the eternal nature not varying from generation and corruption" (1928, p. 231). Plato also posited that the objects of mathematics belong to this eternal realm of being and not the ephemeral worldly realm.

According to Plato, the truths of mathematics are permanent and unchanging. To solve the problem of how humans can come to know about the ideal realm, Plato advanced a theory of reincarnation, claiming that the souls of the unborn dwell in this world of eternal Forms. In the *Meno*, Plato explains that teaching and learning is actually the process of remembering what one already knew before coming into the imperfect worldly existence:

> As the soul is immortal, has been born often, and has seen all things here and in the underworld, there is nothing which it has not learned; so it is no way surprising that it can recollect the things it knew before … nothing prevents a man recovering one thing only—a process men call learning—discovering everything else for himself … for searching and learning are, as a whole, recollection.
>
> *(1981a, p. 70)*

As Norwegian philosopher of mathematics Øystein Linnebo summarized, the *Meno* dialogue reveals two assumptions: that mathematical concepts are innate rather than acquired, and that "mathematical truths are a priori and can be known without relying on experience for one's justification" (2017, p. 6). This is not to say, however, that experience plays no part in the process of remembering mathematical truths. Interestingly, Plato's *Meno* demonstrates that empirical considerations can help individuals come to understand the perfection that is mathematics. In the *Meno*, Socrates uses visual cues to help an enslaved boy gain access to the mathematical realm that is purportedly buried deep inside of the boy from a past life (1981a, p. 70). The allowances made here for the role of sensory experience seem at odds with Plato's assertion in the *Phaedo* that the mind, or soul, can get closer to knowing the ideal realm if it is not deceived by the body (via the senses). Concessions made for sensory experience in the *Meno* may appear to contradict the very economy of Plato's dual-realm model. However, sensory experience functions as a provocation, and a critical one at that. It is an imperfect means of remembering what is true, perfect, and real. "We have no chance of comprehending the Forms without our senses as an intermediary, but we have to be careful to use our intellect to help 'filter' the 'impurities' that our senses may suggest" (K. Parshall, personal correspondence, May 13,

2002). Our array of desert rocks would have been imperfect in its execution, yet the concept of a "perfect square" is indeed perfect.

Many modern-day mathematical Platonists share Plato's belief in the distinction between the imperfect version of mathematical objects that our senses can perceive and the "real" mathematical objects that exist in an abstract realm. However, they depart from Plato by assuming that objects cannot be accessed or understood through our senses. As mentioned earlier, Plato's account of the boy's coming to know certain geometric principles clearly permits (perhaps even requires) employment of the senses. The popular Platonist exclusion of the senses from mathematical understanding, it could be argued, leads to an impoverished Platonism. This hard line can be seen as an example of the broader tendencies at play here, namely that the contemporary Platonist exists in a context influenced by the absolutist–constructivist dualism and, consequently, their ideas tend to be pushed toward extremes. Thus, modern Platonism has turned away from nuance, doubling down on a version that fails to acknowledge how the mathematical realm and our human lives can relate to each other.

The Epistemological Problem of Contemporary Platonism

To the Platonist, mathematical forms are real and are afforded a special ontological status as entities existing in an ideal realm. Mathematical objects, such as the number "2" or the set of all odd numbers, exist in a realm that is neither physical nor mental, as they need no empirical manifestation to be true and, likewise, they need no mind to think of them in order for them to exist (Kitcher, 1983, p. 6). The connection between the knower and what is known, however, remains hazy, and Platonists differ as to how humans come to know elements of the Platonic realm. Recall that Plato's notion was that human souls dwelled in the realm of the Forms prior to birth and need only remember or re-access knowledge of this realm. Modern day Platonists do not typically adhere to this explanation. Kitcher identifies two epistemological positions used by Platonists. The first is intuition[4] and the second is some form of empiricism:

> The former tactic seems to me a desperate measure, tantamount to abandoning the enterprise of explaining our knowledge. The latter requires the Platonist to explain the nature of abstract objects in a way which will enable us to appreciate how standard perceptual processes could furnish information about them.[5] Anti-Platonists are worried by the picture of ethereal entities lurking behind ordinary things, and they wonder how it is possible for the scattering of light from the surfaces of ordinary things to engender knowledge of those entities. The Platonist's task is to provide a better picture.
>
> *(1983, p. 103)*

Indeed, this is no small task. The dilemma, as articulated by Paul Benacerraf in his essay, "Mathematical Truth" (1973), is that Platonists have difficulty explaining how one can possess knowledge of objects that exist in a non-spatiotemporal domain: "The minimal requirement, then, is that a satisfactory account of mathematical truth must be consistent with the possibility that some such truths be knowable" (p. 667). Benacerraf's challenge relates to both the knowability of mathematical truths and to the means by which they are known.

Balaguer's Solution to Platonism's Epistemological Problem

In *Platonism and Anti-Platonism in Mathematics* (1998), Mark Balaguer presents a recent and thorough treatment of platonism. By not capitalizing "platonism," Balaguer is likely acknowledging the differences between the thoughts of Plato and those of contemporary platonists.[6] Balaguer's project is to develop a philosophical justification for platonism and anti-platonism.[7] For our purposes, we will confine ourselves to his case supporting a particular brand of platonism.

Balaguer distinguishes between types of platonism, concerning himself primarily with a "full-blooded" platonism, which he claims "is the only tenable version of platonism" and which capably resolves the problem issued by Benacerraf (p. 21). This full-blooded platonism asserts that every mathematical object that could possibly exist *does* exist. He explains how this is a solution to the epistemological problem by demonstrating how it responds to an argument H. Field (1989) once made about mathematical platonism (as quoted in Balaguer):

> But special "reliability relations" between the mathematical realm and the belief states of mathematicians seems altogether too much to swallow. It is rather as if someone claimed that his or her belief states about the daily happenings in a remote village in Nepal were nearly all disquotationally true, despite the absence of any mechanism to explain the correlation between those belief states and the happenings in the village.
>
> *(p. 49)*

Balaguer admits the failure of the epistemology that would allow for knowledge of a particular Nepalese village without access to it but is quick to point out that not all possible Nepalese villages exist, whereas according to his full-blooded platonism, all possible mathematical objects do exist: "if all possible Nepalese villages existed, then I could have knowledge of these villages, even without access to them. To attain such knowledge, I would merely have to dream up a possible Nepalese village" (p. 49).

One question is whether in adopting his full-blooded brand of platonism, Balaguer is actually talking about mathematics in a fundamentally different way than most platonists. Indeed, at this chapter's outset, we explained that according to our way of characterizing Platonism, it is a philosophy that focuses on the ways in which mathematics exists independently and *externally* to human existence. In his description of what it means to acquire knowledge, Balaguer seems to have shifted from an external to an internal focus for his way of thinking about mathematics:

> If full-blooded platonism is correct, then all consistent purely mathematical theories truly describe some collection of abstract mathematical objects. Thus, to acquire knowledge of mathematical objects, all we need to do is acquire knowledge that some purely mathematical theory is *consistent*. (It doesn't matter how we come up with the theory; some creative mathematician might simply dream it up).
>
> *(Balaguer, p. 48)*[8]

That Balaguer may have crossed the line from platonism to a position based on conditions of consistency is not shocking, as there has been a well-documented historical tendency for mathematical absolutists, when pressed, to shift in this way (Kline, 1980; Hersh, 1997; Parsons, 1967). Platonism, although a widespread way to think about mathematics, cannot resolve some of the philosophical dilemmas it presents. Balaguer's defense of platonism's external body of truths through an emphasis on their internal consistency provides a transition to the other form of mathematical absolutism, namely, formalism.

Formalism

Russell's words in *Mysticism and Logic* might appear to counter an absolutist sensibility: "Mathematics may be defined as the subject in which we never know what we are talking about, nor whether what we are saying is true" (1957, p. 71). Is this the same speaker who extolled mathematics' unique capacity for perfection? The mathematician's divine status? The principles of formalism, as we will see, accommodate this particular brand of "not knowing" yet uphold certainty through their internal stability. While still offering an absolutist account of mathematics in the sense that mathematical truths are viewed as unquestionable once established, formalists tend to regard mathematics as operations that can be performed on sets of symbols. According to most formalists, mathematics bears no necessary relation to the world in which we live (hence, Russell's "what we are talking about"). As Ernest explains:

> The formalist thesis comprises two claims. 1. Pure mathematics can be expressed as uninterpreted formal systems, in which the truths of mathematics are represented by formal theorems. 2. The safety of these formal systems can be demonstrated in terms of their freedom from inconsistency, by means of meta-mathematics.
>
> *(1991, p. 10)*

Unlike Platonism, formalism does not afford a special ontological status to mathematical entities. While the Platonist posits that mathematical objects exist in an ideal realm, the formalist does not go beyond the notion of symbols operating within a given system. Questions about the truth of mathematical statements can therefore only be meaningfully investigated and answered within a well-defined, formal system. Hilary Putnam clarifies this point in a discussion of object and modal descriptions of mathematics (the rough analogs to platonism and formalism, respectively). Of the "object" picture (i.e., platonism), Putnam states: "if one fastens on the first picture, then mathematics is wholly extensional, but presupposes a vast totality of eternal objects." He goes on to say of the modal picture (i.e., formalism): "if one fastens on the second picture, then mathematics has no special objects of its own, but simply tells us what follows from what" (Putnam, 1967, p. 11).[9] The modal description is unconcerned with mathematical "stuff" and instead emphasizes the generating of (presumably sound) deductions. Thus we can see how, to the formalist, it appears that "there are no absolute true mathematical statements," yet true statements exist within and can be generated from the system itself (Ravn & Skovsmose, 2019, p. 94).

The implications of formalism are that mathematics becomes untethered from any physical or Platonic realities, and the focus shifts to internal consistency as a means of legitimacy. A thesis of Morris Kline's *Mathematics: The Loss of Certainty* is that, historically, mathematicians have sought to protect the idea that mathematics is perfect, definitive, and absolute. According to Kline, formalism came about in the wake of alternative mathematical systems that questioned the certainty of the Platonist understanding of the nature of mathematics (1980). While the Platonist places mathematical certainty in the idea that mathematics is above, beyond, or in some way external to the messy uncertainty of any worldly existence, the formalist generally focuses on the certainty of the results that come from the manipulation of mathematical symbols according to formalized rules. Because internal consistency is all that is required, formalism is better able to cope with alternative forms of mathematics, forms which might otherwise threaten to reveal fatal contradictions.

Although Platonist and formalist views about mathematical truth diverge in obvious ways, mathematicians may toggle between them. Hersh describes what he calls the "mathematician's philosophical dilemma":

> The working mathematician is a Platonist on weekdays, a formalist on weekends. On weekdays, when doing mathematics, he's a Platonist, convinced he's dealing with an objective reality whose properties he's trying to determine. On weekends, if challenged to give a philosophical account of this reality, it's easiest to pretend he doesn't believe in it. He plays formalist, and pretends mathematics is a meaningless game.
>
> *(1997, pp. 39–40)*

Parsons, in what is probably a less judgmental piece (Hersh is a vocal opponent of both the Platonist and formalist outlook) makes a similar point: "We begin with Platonism because it is the dominant attitude in the practice of modern mathematicians, although upon reflection, they often disguise this attitude by taking a formalist position" (Parsons, 1967, p. 201). While the Platonist sees no need to justify mathematical pursuits or explain their relevance, as the "realness" of math is evident, formalists argue that in shifting from Platonism to formalism, mathematics does not become devoid of meaning. On the contrary, the formalist would likely say that it is *only* meaningful in relation to the system (Ravn & Skovsmose, 2019).

Perhaps an alternative explanation of the differences between Platonism and formalism will make the characteristics of each more apparent. We have referred to the Platonist's external focus and belief in pre-existing mathematical objects and the formalist's internal focus and tendency to look for truth as the result of following rules. Clearly, a distinction between content (or product) and method (or process) is also at play here. As we have seen, conceiving of Platonism and formalism as tending toward the external and the internal, respectively, can help clarify how each retains its own way of thinking within the larger absolutist "family." Likewise, the product–process dimensions of the two types of absolutism also can help us understand and distinguish them. It should be noted, however, that in the broader philosophy of mathematics conversations it might be more useful to think about both Platonism and formalism as tending toward an external understanding of mathematics, as each presents an understanding of mathematics consisting of elements that are essentially *outside of the sphere of human influence*. This idea gets a bit murkier when the product–process dichotomy is considered, but even here, there is room to configure both the Platonist's products and the formalist's processes as phenomena that are true regardless of whether humans are involved. The conceptualization of both Platonism and formalism as primarily externally focused philosophies will become clearer in the following chapter when absolutism in general will be compared to non-absolutist ways of thinking.

Formalism's Origins

German mathematician and philosopher Gottlob Frege is commonly considered the modern father of mathematical logic. Drawing inspiration from Leibniz's notion that mathematics ultimately comes from logic and the Leibnizian belief that human reasoning can be reduced to a formal symbolic system, Frege set out to define arithmetic and the concept of number in terms of logic. He contended that mathematics is analytic, meaning that the truths of mathematics follow from their logical premises. Kline states the Fregean point of view:

They (the laws of mathematics) say no more than what is implicit in the principles of logic, which are a priori truths. Mathematical theorems and their proofs show us what is implicit. Not all of mathematics may be applicable to the physical world but certainly it consists of truths of reason.

(Kline, 1980, p. 217)

Frege's attempt to derive mathematics from logic failed, as just prior to the publication of his work, Russell noted a potential contradiction in Frege's concept of "the set of all sets." Nonetheless, as Kline has noted, reducing mathematics to a formalized system remained an attractive idea to philosophers of mathematics wanting to ensure the certainty of the discipline. The failure of Frege did not convince Russell that searching for certainty in mathematics by means of the mathematics–logic connection was a fruitless effort.[10]

Russell and Whitehead's Principia Mathematica

The logicism of Bertrand Russell and A.W. Whitehead is frequently considered a freestanding philosophy of mathematics (Ernest, 1991; Stroll, 1999, p. 610), but it can also be considered a version of formalism. According to the logicists, mathematics is a branch of logic and all mathematics is reducible to logical notions. Whitehead and Russell's *Principia Mathematica* is widely considered the clearest example of the logicist thesis (Stroll, 1999; Ernest, 1991) and an attempt to shore up the foundations of mathematics through an increased axiomatization. While ultimately doomed to failure, the ideas put forth in the *Principia* were and still are influential in philosophy of mathematics, logic, and even computer science. We are interested, of course, in exactly how Russell and Whitehead's logicism is classifiable as a formalist endeavor.

In *Portraits from memory* (1956), Russell explains the rationale for his work:

> I wanted certainty in the kind of way in which people want religious faith. I thought that certainty is more likely found in mathematics than elsewhere. But I discovered that many mathematical demonstrations, which my teachers expected me to accept, were full of fallacies, and that, if certainty were indeed discoverable in mathematics, it would be in a new field of mathematics, with more solid foundations than those that had hitherto been thought secure.

(p. 54)

In *Introduction to Mathematical Philosophy* (1919/1963), a sketch of the full-blown project detailed in the *Principia*, Russell wrote explicitly of the relationship between logic and mathematics:

> it has now become wholly impossible to draw a line between the two; in fact, *the two are one*. They differ as boy and man: logic is the youth of mathematics and mathematics is the manhood of logic.

(p. 194, emphasis added)

To support this connection, Russell explained that beginning with logical premises and using deduction to finish with products that are mathematical, there is "no point at which a sharp line can be drawn" (1919/1963, p. 194).[11]

Russell and Whitehead first took the time to develop logic, per se, and its fundamental concepts (proposition, propositional functions, quantifiers, etc.). Not unlike the structure of Euclidean mathematics, with its axiomatic economy and rules of inferences, the theorems in the *Principia* were deduced from established constituent parts. But the search for certainty was still threatened by paradox, so Russell and Whitehead worked to find a way around the troubles that befell Frege's concept of the set of all sets. Essentially, they needed to avoid the possible contradictions that follow from the inclusion of a set as an instance of itself. A clear explanation of this contradiction, as well as Russell and Whitehead's way out of the dilemma (the theory of types) is provided by Kline's summary of an example from the *Principia*:

> Let us consider the contradiction posed by the statement "all rules have exceptions" (Chapter IX). This statement is about particular rules such as "all books have misprints." Whereas the statement about all rules is usually interpreted to apply to itself and so leads to the contradiction that there are rules without exceptions, in the theory of types the general rule is of higher type and what it says about particular rules cannot be applied to itself. Hence, the general rules need not have exceptions.
>
> *(Kline, 1980, p. 223)*

By establishing separate levels of sets, Russell and Whitehead attempted to sidestep the problems of the sort detailed above (i.e., that if all rules have exceptions, and the prior statement is a rule, then there must be exceptions to it, hence if true it is false). At least in terms of the *set of all sets* problem, Russell and Whitehead's system did free itself of contradictions. The question, however, is at what price?

The inclusion of the theory of types was quite complex and required that Russell and Whitehead introduce several other axioms as well, such as *reducibility*, the existence of *infinite classes*, and the *axiom of choice* (Kline, p. 223). Philosophers of mathematics reacted against the project on different grounds. There were questions of whether axioms were really pure logic. If they were not, then it would seem that Russell and Whitehead's work, while possibly free of inconsistencies, was simply one formalized system that happened to work and not proof of the certainty of mathematics. The authors gradually backpedaled from their initial findings. Whitehead did not even have his name included in the third edition of the *Principia*. Russell was not so easily convinced about the futility of his project, but he gradually drifted from his original thought until he had more or less reversed his stance on what he had attempted to accomplish. As Russell said of the entire enterprise of fortifying the foundations of mathematics:

as the work proceeded, I was continually reminded of the fable about the elephant and the tortoise. Having constructed an elephant upon which the mathematical world could rest, I found the elephant tottering, and proceeded to construct a tortoise to keep the elephant from falling. But the tortoise was no more secure than the elephant, and after some twenty years of very arduous toil, I came to the conclusion that there was nothing more that I could do in the way of making mathematical knowledge indubitable.

(Russell, 1956, pp. 54–55)

While Russell and Frege's projects of reducing mathematics to formal logic have been abandoned by the vast majority of philosophers of mathematics, Benacerraf and Putnam (1964) detail the lasting impact of logicism:

Yet, it should not be forgotten that if today it seems somewhat arbitrary just where one draws the line between logic and mathematics, this is itself a victory for Frege, Russell, and Whitehead: before their work, the gulf between the two subjects seemed absolute.[12]

(p. 11)

Hilbert's Project

David Hilbert's formalism grew out of a dislike for other projects that attempted to shore up the foundations of mathematics (Hersh, 1997; Kline, 1980; Benacerraf & Putnam, 1964, p. 6). Hilbert was disturbed by Russell and Whitehead's logicism. He was equally troubled by the philosophical movement of intuitionism, particularly the work of Luitjens Brouwer. Brouwer proposed that humans unexplainably can come to know the natural numbers, all other mathematics comes about after this "intuition," and any supposed mathematical objects that did not come from the natural numbers are not and cannot be found to be meaningful (Kline, 1980; Gowers, 2008, p. 799).[13]

Hilbert was unconvinced by efforts to explain the certainty of mathematics by either reducing it to logic or by attributing to humans the mysterious ability of some form of primal intuition about the nature of number. He, along with many others, was also concerned about the effect of the concept of infinity on the certainty of mathematics. In fact, the role of the infinite in helping or hindering our understanding of the nature of mathematics was a hot topic of the day for virtually all involved. As Benacerraf and Putnam explain:

Hilbert argues convincingly that physics provides no clear evidence for the existence of such structures: in fact, the progress of physics has, as he points out, introduced finiteness and discontinuity in area after area in which the infinite and continuous once reigned supreme.

(1964, p. 6)

Hilbert, who is credited with developing a way of thinking about mathematics that loosened its connections to the empirical world, opted to appropriate physics for his description of why the infinite had no place in the foundations of mathematics. Since the infinite was a cause for doubt about the certainty of mathematics, Hilbert worked to construct a system that could stand on its own without needing the concept of infinity as a precondition. The doubt that physics cast upon the concept of infinity obviously suited his purpose, which was "to establish once and for all the certitude of mathematical methods" (Hilbert, 1964, p. 135). Hilbert further explained the need for some means of coping with the nettlesome problem of infinity: "But a still more general perspective is relevant for clarifying the concept of the infinite. A careful reader will find that the literature of mathematics is glutted with inanities and absurdities which have had their source in the infinite" (p. 135).

In a paper delivered to the Hamburg Mathematical Seminar in 1927 and reprinted as "The Foundations of Mathematics," Hilbert stated his goals and a rough idea of how he planned to accomplish them:

> With this new way of providing a foundation for mathematics, which we may appropriately call a proof theory, I pursue a significant goal, for I should like to eliminate once and for all the questions regarding the foundations of mathematics, in the form in which they are now posed, by turning every mathematical proposition into a formula that can be concretely exhibited and strictly derived, thus recasting mathematical definitions and inferences in such a way that they are unshakeable and yet provide an adequate picture of the whole science. I believe that I can attain this goal completely with my proof theory, even if a great deal of work must still be done before it is fully developed.
>
> *(Hilbert, 1967, p. 465)*

And so Hilbert constructed a system that sought to utilize only provable components that yielded no contradictions. These two ideas were known as consistency and completeness (Bernays, 1967, p. 501). To do so he created a formal system of symbols that did not employ natural language as previous systems of axiomatization had. Furthermore, Hilbert believed that his system was a success in that it established that working with formal rules within his closed system, one could not produce contradictory results (e.g., that $0 = 1$).

A Fregean Response to Hilbert

Frege, a forerunner to Russell and Whitehead, criticized Hilbert's attempt at formalization. He argued that the development of a consistent system was no remarkable accomplishment, as Hilbert's system did not refer to anything outside its own boundaries. In "Frege's New Science" (2000), Aldo Antonelli and Robert May explain this critique:

> Hilbert saw this result as *mathematically* significant because for him this constituted a consistency proof. But to Frege its significance could only be merely formal, for all that would have been done is attribute to sentences a property of propositions. It is *propositions*, not their formal guises, that are consistent or not, and this is not something that requires proof; if all thoughts of a set are true, they must be consistent.
>
> *(p. 264)*

This Fregean critique did not deal too serious a blow to Hilbert's project. Frege did not so much doubt Hilbert's conclusions, instead the two men disagreed about the significance of what Hilbert had accomplished. Kurt Gödel's incompleteness theorem cast more serious doubt upon the worth of Hilbert's project. His theorem basically states that any reasonably complex formal system will be incomplete or will contain inconsistencies. As Kitcher restates: "For any formal system which is rich enough to be a candidate for stating all mathematical truths, there will be a true sentence in the language of the system which is not a theorem of the system" (1983, p. 139).

It is commonly held that Gödel's theorem dashed the possibilities of Hilbert successfully completing his vision of eliminating uncertainty within the foundations of mathematics. This is not to say, however, that Hilbert's work was in vain. His formalization of mathematical proof has served to stimulate new areas of development within mathematics.[14]

Absolutist Mathematics and Rorty's Distinction

In Chapter 1, we introduced Rorty's characterization of the deep divide between philosophers who think of reality as primarily external, independent of human existence, and fundamentally "found," and philosophers who emphasize the manner in which humans construct or "make" their own reality. It is not, we should think, a stretch to see the tension between the Platonist's depiction of mathematics as completely external and free of human involvement and the formalist's conception of mathematics as existing entirely within the boundaries of a closed system as an instance of Rorty's distinction.

According to this conception, the Platonist can only hope to come to know or to *find* preexistent mathematical objects. Conversely, the formalist creates or *makes* formal systems that are legitimate by means of their internal consistency. As we earlier suggested, this version of the found–made distinction makes sense within the context of mathematical absolutism. However, when viewed from a perspective other than an absolutist one, both Platonism and formalism can be seen as each representing the *found* side of the distinction. The key to understanding this shift is viewing the formalist's task not so much as constructing or making mathematics as it is to discover the logical relations that allow a formal system to work. Thought of in this way, formalism is merely a different type of search for a mathematics that is made up of a priori and extra-human (i.e., logical) phenomena. So in the context of broader discussions within the philosophy of mathematics, it is frequently the case that both Platonists and formalists are understood as trying to find and not make mathematics.

A few more words about the neatness with which we have presented this characterization are in order. At the outset, we called absolutist perspectives a "family of related approaches" and explained that the work of all of the thinkers presented is far too complex to neatly fit into any category. Still, we hoped that our characterizations would shed light on the different ways of thinking about mathematics as permanent and unchanging. Figure 3.2 depicts the major categories within absolutism in relation to one another. While we categorized each philosopher's work (Balaguer's platonism, Russell's logicism/formalism, Hilbert's formalism, etc.), we noted that Balaguer's full-blooded Platonism might actually blur the lines between Platonism and formalism. A case could be made that most strains of formalism that were presented operate with some form of latent Platonism lurking in the shadows. In "Foundations of Mathematics," Parsons includes a discussion of how Frege, Russell, Cantor, and Hilbert all used formalism to support what is essentially a Platonist understanding of mathematics: "where both number and sets or sequences of numbers were treated as existing in themselves" (Parsons, 1967, p. 189).[15] In this way, the formalist movement can be interpreted as a radical shift to an emphasis on the consistency of insular, man-

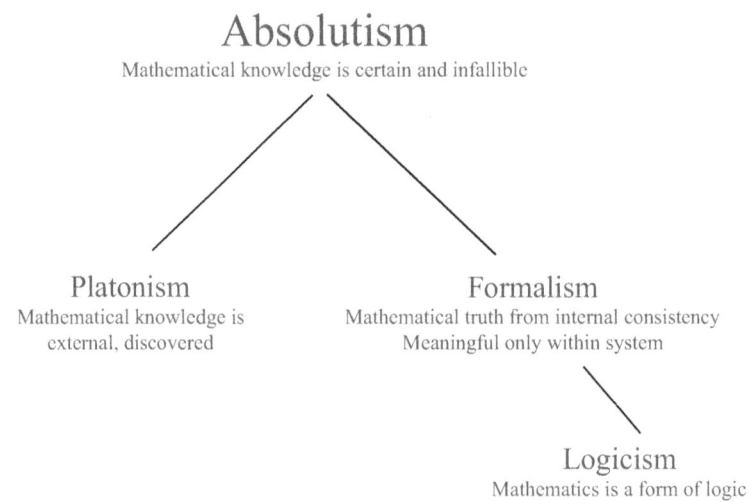

FIGURE 3.2 The Absolutism Family

made systems in an effort to prove the existence of a completely external, pre-existent body of truths. While this might seem perplexing, it is helpful to recall Morris Kline's thesis that mathematicians and philosophers of mathematics were not willing to give up the certainty of mathematics and, in the face of what were perceived as the unsettling implications of the development of some new types of mathematics, they turned to formalism in hopes of preserving the indubitability of mathematics. The paradoxes that vexed Frege, Russell, and Hilbert could not and did not shake loose the hold of absolutism.

The absolutist orientation has proven quite resilient and it has endured, in part, because it is useful in some regards. For example, Platonism allows mathematicians to adopt a natural scientist-like discovery mode. Formalism provides a respectable retreat for mathematicians when questioned about just how real Platonic math truths actually are, not to mention the very concrete uses of formalism in computer programming and artificial intelligence. However, absolutism as an overarching philosophy of mathematics is a failure.[16] Conceiving of the mathematician's job as trying to get in touch with some extra-human body of truths or as trying to develop internally consistent systems of thought clearly is not an adequate explanation of what it is that mathematicians actually do, as mathematicians (and everyday users of mathematics, for that matter) do more with mathematics than simply discover it. Absolutism tends not to offer any explanation of how humans can come to know and use mathematics in any significant sense. In this regard, even on its own terms, absolutism is a failure, as the Platonist is unable to account for how we access the Platonic realm and formalists are unable to explain how it is that mathematics helps to solve real-world problems outside of the context of particular mathematical systems. Furthermore, according to absolutism, there is little appreciation of the role that humans play in creating mathematics. Finally, absolutism as a philosophy of mathematics denies the existence of any elements of mathematics that are contingent upon relations with the impermanent empirical world in which humans dwell.

In the chapter that follows, we consider the ideas of some non-absolutist philosophers of mathematics. Whether these opponents of mathematical absolutism can develop an alternative way of thinking that avoids the dogmatic tendencies suggested by absolutism is a question that will be considered throughout the next chapter. Interestingly, in spite of the fundamentally different ways in which non-absolutist philosophers tend to conceive of mathematics, we contend that an internal–external divide that is roughly analogous to the split within the absolutist perspective persists.

Notes

1 Here, we capitalize the second "Truths" as a nod to Euclid's five fundamental axioms, the unproven but uncontestable assumptions about construction, congruence, and parallelism from which all subsequent theorems were derived (Ravn & Skovsmose, 2019).

2 While some might argue that Descartes's recognition that new knowledge is possible, as well as his encouraging words about thinking for oneself at the outset of *Discourse on*

Method, would place him outside the pale of absolutism, Ernest refers to the Cartesian version of knowledge as modeled on axiomatic proof as his rationale for placing Descartes with the absolutists. Additionally, Dossey (1992) explains Descartes's project as working to "separate it [mathematics] from the senses" (p. 40).

3   A standard definition of a priori, according to Runes's *Dictionary of Philosophy*, is the Kantian notion of an entity that is valid "independent(ly) of all impressions of sense ... and unmixed with anything empirical." Runes explains a broader understanding of a priori that includes "anything non-empirical, or something which can be known by reason alone" (Runes, 1942, p. 16).

4   This is not to be confused with the intuitionism of L.E.J. Brouwer.

5   See our earlier discussion on the *Meno*.

6   Following Balaguer, we have elected not to capitalize "platonism" for the remainder of the book, except in instances where the ideas are proximal to those of Plato himself.

7   His interesting conclusion is that because we could never determine whether or not mathematical objects exist, there is no "fact of the matter" regarding their existence.

8   This discussion provides a good opportunity to observe how the dualisms of today's context can distort ideas. Plato's thinking existed in a time during which the forced choice binaries were less present. Today's Platonists have to react to and defend against thinkers on the "other side" of a number of divides (found/made, internal/external, and so on).

9   A divergent account of the formalist ontology is that the "realness" of mathematics is not found in an abstract realm, but rather in the act of proof. In *Hilbert's Program*, Detlefsen calls Hilbertian formalism a "modified realism." He goes on to say that "according to this view, the ontological commitments are located not in those parts of mathematics which we use to acquire knowledge but rather in those propositions which are used to establish the reliability of the mathematics thus used" (1986, p. 3).

10 Of course, there are other notable formalists. For example, Ravn and Skovsmose give substantial credit to American mathematician and logician Haskell B. Curry (1900–1982) for the development and advancement of formalism. It was Curry's core thesis, a rejection of Platonistic perspectives, that led him to conclude that "a formalization is a necessary condition for turning philosophical investigations into meaningful investigations" (Ravn & Skovsmose, 2019, p. 92). Curry adamantly dismissed the notion of eternal, absolute mathematical truths.

11 As we all well know, no one in their right mind would actually read the *Principia* (and we are no exception). The publishers were disturbed by the cost and labor required to produce a book that developed a new symbology and its generally imposing nature has put off generations of would-be readers. This very brief sketch of the *Principia* came about from C.I. Lewis's review of the work (1914) and also from Kline (1980).

12 Neo-Fregeanism is a movement that resurrected something resembling logicism. Crispin Wright's *Frege's Conception of Numbers as Objects* (1983) sparked an interest in the reconsideration of Frege's work. See Arché Research Project's *The Logical and Metaphysical Foundations of Classical Mathematics* website (https://web.archive.org/web/20070209123549/http://www.st-andrews.ac.uk/~arche/pages/projects/mathsproject.html) for more on this phenomenon.

13 Note the similarities between Brouwer's ideas and Kitcher's discussion of the epistemological dilemma faced by Platonists' attempts to explain how we come to know mathematics.

14 Hilbert's confidence in the infallibility of a carefully defined mathematical system followed him quite literally to the grave. Inscribed on his tombstone is his famous rejoinder, *Wir müssen wissen. Wir werden missen.* ("We must know. We will know.") We might say that formalism's attention to rigor anticipated the work of computer programmers, who labor under the specter of problems they cannot foresee. They must construct a logical apparatus that, they hope, will return consistently sound output given any input.

15 Parsons's assertion regarding the latent Platonism operating behind the formalists helps explain why he chooses to view Platonism as the overarching form of absolutism and he relegates formalism to a subgenre.

16 While absolutism does not do justice to the pragmatic, dynamic, and creative facets of mathematics, it is also possible that absolutism fails as an "overarching philosophy of mathematics" because seeking to articulate "overarching philosophies of mathematics" is not a useful endeavor. We are not arguing that thinking deeply about the nature of mathematics is without worth, but that there is trouble in doing so with the aim of reaching a final, absolute conclusion. The pragmatic/evolutionary philosophy of mathematics that we develop in later chapters is offered in the spirit of providing a useful way of thinking about mathematics given the tasks of teaching and learning it and given that schooling exists, in part, to foster democratic participation in wider society.

References

Antonelli, A., & May, R. (2000). Frege's new science. *Notre Dame Journal of Formal Logic*, 41(3), 242–270.

Balaguer, M. (1998). *Platonism and anti-platonism in mathematics*. New York: Oxford University Press.

Benacerraf, P. (1973). Mathematical truth. *The Journal of Philosophy*, 70(19), 661–679.

Benacerraf, P., & Putnam, H. (1964). *Philosophy of mathematics: Selected readings*. Englewood Cliffs, NJ: Prentice-Hall.

Bernays, P. (1967). David Hilbert. In P. Edwards (Ed.), *The encyclopedia of philosophy* (pp. 496–526). New York: The Macmillan Company and The Free Press.

Detlefsen, M. (1986). *Hilbert's program: An essay on mathematical instrumentalism*. Boston: D. Reidel.

Dossey, J. (1992). The nature of mathematics: Its role and its influence. In D.A. Grouws (Ed.), *Handbook of research on mathematics teaching and learning: A project of the National Council of Teachers of Mathematics* (pp. 39–48). New York: Macmillan.

Ernest, P. (1991). *The philosophy of mathematics education*. Bristol, PA: The Falmer Press.

Ernest, P. (1998). *Social constructivism as a philosophy of mathematics*. Albany, NY: State University of New York Press.

Field, H. (1989). *Realism, mathematics, and modality*. Oxford: Basil Blackwell.

Frege, G. (1974). *Foundations of arithmetic* (J. Austin, Trans.). Evanston, IL: Northwestern University Press.

Gowers, T. (Ed.) (2008). *The Princeton companion to mathematics*. Princeton, NJ: Princeton University Press.

Hardy, G. (1940/2001). *A mathematician's apology*. Cambridge: Cambridge University Press.

Hersh, R. (1997). *What is mathematics, really?* New York: Oxford University Press.

Hilbert, D. (1964). On the infinite. In P. Benacerraf & H. Putnam (Eds.), *Philosophy of mathematics: Selected readings* (pp. 134–151). Englewood Cliffs, NJ: Prentice-Hall.

Hilbert, D. (1967). The foundations of mathematics. In J. van Heijenoort (Ed.), *From Frege to Gödel: A source book in mathematical logic, 1879–1931* (pp. 464–479). Cambridge, MA: Harvard University Press.

Kitcher, P. (1983). *The nature of mathematical knowledge*. New York: Oxford University Press.

Kline, M. (1980). *Mathematics: The loss of certainty*. New York: Oxford University Press.

Lewis, C.I. (1914). Review of A.N. Whitehead and Bertrand Russell, *Principia Mathematica*. *Journal of Philosophy*, 11(18), 497–502.

Linnebo, Ø. (2017). *Philosophy of mathematics*. Princeton, NJ: Princeton University Press.

Parsons, C. (1967). Foundations of mathematics. In P. Edwards (Ed.), *The encyclopedia of philosophy: Vol. 5* (pp. 188–213). New York: The Macmillan Company and The Free Press.

Plato (1928). *The republic* (B. Jowett, Trans.). New York: Charles Scribner's Sons.

Plato (1981a). Meno. In *Five dialogues: Euthyphro, Apology, Crito, Meno, Phaedo* (G.M.A. Grube, Trans.). Indianapolis, IN: Hackett.

Plato (1981b). Phaedo. In *Five dialogues: Euthyphro, Apology, Crito, Meno, Phaedo* (G.M.A. Grube, Trans.). Indianapolis, IN: Hackett.

Press, G. (1999). Plato. In R. Popkin (Ed.), *The Columbia history of western philosophy* (pp. 32–51). New York: MJF Books.

Putnam, H. (1967). Mathematics without foundations. *The Journal of Philosophy*, 64(1), 5–22.

Quinton, A. (1967). Knowledge and belief. In P. Edwards (Ed.), *The encyclopedia of philosophy: Vol. 4* (pp. 345–352). New York: The Macmillan Company and The Free Press.

Ravn, O., & Skovsmose, O. (2019). *Connecting humans to equations: A reinterpretation of the philosophy of mathematics*. New York: Springer.

Reid, C. (1996). *Julia: A life in mathematics*. Washington, D.C.: Mathematical Association of America.

Rorty, R. (1999). *Philosophy and social hope*. New York: Penguin Books.

Runes, D. (1942). *The dictionary of philosophy*. New York: Philosophical Library.

Russell, B. (1919/1963). *Introduction to mathematical philosophy*. Edinburgh, Scotland: Neill & Co. Ltd.

Russell, B. (1956). *Portraits from memory*. New York: Simon and Schuster.

Russell, B. (1957). *Mysticism and logic*. Garden City, NY: Doubleday Anchor Books.

Russell, B. (2002). *The selected letters of Bertrand Russell, volume 1: The private years 1884–1914*. (2nd ed., N. Griffin, Ed.). New York: Routledge.

Stroll, A. (1999). Twentieth-century analytic philosophy. In R. Popkin (Ed.), *The Columbia history of western philosophy* (pp. 604–666). New York: MJF Books.

4

NON-ABSOLUTIST MATHEMATICS

The Constructivist Thesis

$$\frac{3}{4} = 0.34 \qquad\qquad \frac{2}{31} = 0.21 \qquad\qquad \frac{7}{90} = 0.70$$

Though puzzling, these equations, which were written with confidence and conviction by an eighth grader, have a rhyme and reason. They reflect an invented strategy for writing the decimal equivalents of fractions, a strategy that was based on examples provided by a teacher:

$$\frac{1}{10} = 0.10 \qquad\qquad\qquad \frac{2}{10} = 0.20$$

Notice that in each of the teacher's decimal expressions, the digit in the tenths place corresponds to the fraction's numerator, just as the digit in the hundredths place seems to be taken directly from the ones place of the fraction's denominator. Hence, to the student, $2/31 = 0.21$. At first glance, the student's decimal equivalents may seem preposterous to us. After all, 3/4 is greater than one half, so how could its decimal expression possibly be *less* than 0.50? Yet to the student it was a perfectly sensible answer.[1] There is a kind of ingenuity in the way the student extracted a pattern and forged a strategy, one that would be of service until a counterexample (say, $25/100 = 0.25$) demanded some kind of retooling, or rule revision.

As is the case with absolutist philosophies of mathematics, non-absolutism also takes many forms. As the name suggests, the most common unifying characteristic of non-absolutist philosophies of mathematics is their opposition to absolutist perspectives (Bickhard, 1998, p. 99). Absolutism's role as a catalyst in the development of non-absolutist conceptions of mathematics is captured nicely by philosopher of mathematics education Paul Ernest: "The absolutist view of mathematical knowledge has been subject to a severe, and in my view, irrefutable criticism. Its rejection leads to the acceptance of the opposing fallibilist view of mathematical knowledge" (1991, p. 18). If rejecting absolutism is tantamount to rejecting the certainty of mathematical truth, then this is a "truth" now fallible and subject to correction.

The most influential forms of non-absolutism (both within and beyond the boundaries of what is considered mathematics) tend to cluster under the general banner of "constructivism." But this banner has been run up the flagpole in support of so many different and divergent causes that at the outset of the volume, *Constructivism in Education*, Denis Phillips questions whether labeling someone a constructivist "tell(s) us anything clearcut about his or her position," and declares that embarking on this study of constructivism is "the beginning of a trip into a nightmarish landscape" (2000, p. 7).

We find "nightmarish" to be something of an exaggeration. Still, the landscape we traverse in this chapter is indeed complex and difficult. We begin with Catherine Fosnot's general notion of constructivism from *Constructivism: Theory, Perspectives and Practice*. It is worth quoting at length, as it foreshadows some of the crosscurrents and controversies within the movement that will be considered in this chapter:

> Constructivism is a theory about knowledge and learning; it describes both what "knowing" is and how one "comes to know." Based on work in psychology, philosophy, and anthropology, the theory describes knowledge as temporary, developmental, nonobjective, internally constructed, and socially and culturally mediated.
>
> *(Fosnot, 1996, p. ix)*

She goes on to explain that learning, to the constructivist, can be seen as a

> self-regulatory process of struggling with the conflict between existing personal models of the world and discrepant new insights, constructing new representations and models of reality as a human meaning-making venture with culturally developed tools and symbols, and further negotiating such meaning through cooperative social activity, discourse, and debate.
>
> *(Fosnot, 1996, p. ix)*

Fosnot proclaims that constructivism is not a theory of teaching but that it does "suggest" major departures from traditional forms of instruction. This might be the first glimpse at Phillips's "nightmarish landscape," as separating the pedagogical, epistemological, and possibly even the ontological facets of this way of thinking is quite complex. For example, while Fosnot describes a fairly clear line between constructivism and theories of pedagogy, other commentators do not.

In their acclaimed "Young Mathematicians at Work" series, Fosnot and Dolk do not contradict Fosnot's earlier distinction, but they intentionally blur the line between teaching and learning, explaining how various cultures have different ways of thinking about the relationship between teaching and learning and that language differences can shed light on these relationships. They point out that some languages do not have separate words for teaching and learning and they use Dutch as an example, explaining that "the distinction between teaching and learning is made only by the preposition. The verb is the same. *Leren aan* means teaching; *leren van* means learning" (Fosnot and Dolk, 2001, p. 1). Fosnot and Dolk argue that the linguistic closeness between teaching and learning suggests

> an integration in learning/teaching frameworks: teaching will be seen as closely related to learning, not only in language and thought but also in action. If learning doesn't happen, there has been no teaching. The actions of learning and teaching are inseparable.
>
> *(Fosnot and Dolk, 2001, p. 1)*

The authors make no claim that constructivism is a theory of teaching, per se. They do, however, argue that math teacher education should be rooted in constructivism and phronesis, an Aristotelian term that loosely translates to practical wisdom, with an overarching goal of helping current or future teachers "develop a new conception of the nature of mathematics, one based on the human activity of making meaning through a mathematical lens" (p. 173). In other words, rather than proselytize constructivist learning theory explicitly in the hopes that its core tenets will somehow manifest in teachers' future practice, they argue that teachers of teachers should design experiences in which participants themselves engage in mathematical meaning-making and construct "situation-specific" pedagogical knowledge. Constructivism remains a theory of learning here, but what is potentially under construction is a new or revised conception of teaching.

By contrast, Michael Matthews—not entirely approvingly—has called constructivism "education's version of a grand unified theory" (Matthews, 2000, p. 161). In the same passage he explicitly states that constructivism *is* "a theory of teaching, a theory of education, a theory of educational administration, a theory of the origin of ideas, a theory of both personal and scientific knowledge and even a metaphysical and ideological position" (p. 161). If this chapter is going to avoid becoming part of Phillips's "nightmarish landscape" we will need to proceed carefully. Determining what type of constructivism and whether we are even talking about constructivism can be difficult, as we have already seen in the Fosnot/Matthews disagreement.

Philosophically too, problems arise from blending flavors and strains of constructivism (Phillips, 1987, pp. 137–157). For example, constructivist epistemology and constructivist theories of pedagogy are certainly not the same thing and yet it seems that frequently there is confusion in this regard. In "Constructing Constructivism, Epistemological and Pedagogical," Howe and Berv (2000) discuss what they call the "looseness of fit" between constructivism as epistemology and constructivism as a theory of pedagogy (p. 19). While we aim to avoid the type of conflations that we (and Howe and Berv) have warned about, we may need to intentionally blur the lines between philosophy, mathematics, and education. Some interplay is not just unavoidable, it is in fact desirable and sometimes necessary to capture the richness and complexity of how students come to think of mathematics.

Still more potential confusion around constructivism comes from the fact that, with a few notable exceptions, the teaching and learning of mathematics are frequently overlooked by philosophers of mathematics.[2] Later we confront this situation more directly, but for now it is useful to keep these interdisciplinary relations (or lack thereof) in mind. The challenge lies in trying to be as clear as possible about where ideas are coming from and to remember that the neat lines that have been drawn for the purpose of clarity have been drawn for the sake of this project and are not part of the landscape, nightmarish or otherwise, in any tangible sense.[3]

What follows is a historical account of the roots of the constructivist movement. We characterize contemporary constructivism as possessing two main branches, psychological and social, and use the work of several thinkers to help to clarify the psychological and social aspects of constructivism as well as constructivism in general.[4] We engage with and critique both versions of constructivism, setting the stage for a proposed alternative designed to contend with problems that emerge from adopting strictly absolutist or constructivist conceptions of mathematics.

The Roots of Constructivism: From Locke and Descartes to Kant and Beyond

Ernst von Glasersfeld states that all constructivists agree with the following idea: "The notion that knowledge is the result of a learner's activity rather than of the passive reception of information or instruction" (1991, p. xiv). This basic tenet of constructivist thought has been present at least since the time of ancient Greece—recall the Socratic insistence on dialogue as a means to help learners build new understandings, as opposed to more direct, didactic notions of learning. The notion of active learning is a generally accepted truth, so much so that it alone is not a bold enough declaration to establish a different school of thought than mainstream epistemology. As Howe and Berv explain: "constructivism must pick out something deeper than this, something deep enough to distinguish among epistemological views. Otherwise, it may as well be dropped as superfluous" (2000, p. 20).

Many who call themselves constructivists do develop a "deeper" theory that is epistemologically distinct, sometimes even threatening to others. In general, the deeper, more controversial claims of constructivists break into two clusters. The constructivists with a focus on individual ways of knowing put forward a version that casts doubt on the ability of the objective world—to the extent that there even is one, according to this way of thinking—to offer any insight into reality. To many individual-focused constructivists, the understandings that we come to might be wrong and regardless of their verity, they are not verifiable according to any external standard. Socially-minded constructivists have a tendency to cast doubt upon the traditional idea that humans, through science, are converging upon Truth, or even a greater understanding of some objective reality.[5]

Continuing our brief history of constructivism, we now jump forward to the empiricism of John Locke and the rationalism of René Descartes. Both are described as advocating "non-(or half-way) constructivist epistemologies" (Howe & Berv, 2000, p. 20). Locke's empiricism offers a very passive version of knowledge acquisition, as sensory input impresses itself on the mind (Locke, 1979). The only active or constructive element in his brand of empiricism is when, after the fact, the mind orders and classifies "what is already *given* in experience" (Howe and Berv, 2000, p. 20). The rationalism of Descartes, while sharing certain commonalities with some of the more radical flavors of contemporary constructivism, at least in terms of the boundedness of the mind of the individual knower, is considered only partial constructivism in that it is fixed truths that individual Cartesian minds can come to know through rational ideas and the deductive process (Descartes, 1968). Phillips uses Rodin's statue *The Thinker* as a metaphor for the way in which Cartesian rationalism works: "the Thinker is a solitary figure, deeply engrossed in cogitating about the world's problems, using nothing but the power of his rational intellect" (Phillips, 2000, p. vii). Phillips calls Descartes's *cogito ergo sum* "the *indubitable* foundation upon which he erected all the rest of his beliefs" (p. vii, emphasis added). Among these beliefs was the certainty of truths based on axioms and derived through intuition and deduction. While for Descartes it did take mental activity to make knowledge possible, it was still an uncovering of a preexistent, *indubitable* type of knowledge.[6]

Immanuel Kant is commonly considered the first major philosopher to forward what can reasonably be considered a constructivist outlook, as he attempted a synthesis of the two "not-quite" constructivisms of rationalism and empiricism. Kant disagreed with the empiricists, arguing that there can be no sensory input prior to mental manipulation and disagreed with the rationalists, arguing that mental content (excluding, of course, a priori mental structures) is not independent of experience: "Whenever we think of objects, we have to use sensibility, intellect, and apperception" (Kant, 1963 p. 130). Kant also posited that objective reality, his "things in themselves," cannot be known by us *because of* our process of constructing, or as Bryan Magee states, our "apparatus for experiencing" such objects (Magee, 1997, p. 143).

Kant was not completely subjectivist when it came to the human construction of knowledge. While it is true that, to Kant, the mind's categories help to organize sense data in each individual, the structures are themselves universal. Temporality, space, causation, etc. are present in all minds and stable from individual to individual. Howe and Berv call this commonality of the elements of construction, *intersubjectivism*. So, while it is not difficult to recognize the Kantian synthesis of rationalism and empiricism as an explicitly constructivist epistemology, there were very definite limits to the nature of these constructions, namely the sensory input and the categories of the mind.

Post-Kantian constructivism branched out in many directions. Howe and Berv postulate that 20th-century philosophy replaced Kant's transcendental categories with language, as language presumably constrains the ways in which we can come to know or construct our worlds. In this regard, Wittgenstein's work is a clear example of post-Kantian constructivism. Likewise, Piaget's brand of constructivism, with its emphasis on how individuals actively construct knowledge through mental sorting and structuring, has much in common with Kant's, so much so that Piaget frequently referred to himself as a neo-Kantian. By other accounts, Marx's materialism is a form of constructivism, in that groups or classes of people construct their understanding of their world (or have it constructed for them) based on their material circumstances. Marx influenced Vygotsky, who in turn focused on how social relationships aided in the construction of knowledge.

This brief sketch of the development of some strains of constructivist thinking is not intended to be complete. It does, however, set the stage for what follows in the rest of this chapter, including a presentation of Piaget's work as inspiring a more radical psychological constructivism, and Vygotsky's work as a precursor to more radical forms of social constructivism. We acknowledge that Chapter 4 is considerably longer than Chapter 3 and takes an admittedly more tortuous path to its destination. This is no accident: absolutist versions of mathematics tend to restrict input from outside the boundaries around mathematical understanding, whereas constructivism invites, or at least allows, contributions to mathematics from other disciplines and a variety of types of knowledge. As a result, non-absolutist notions of mathematics are inherently messier than absolutist versions and, consequently, there are challenges that come with presenting a clear picture of mathematics from non-absolutist, fallibilist perspectives. The terrain of constructivism is vast and complex. Figure 4.1 is a (deceptively, we admit) simplified

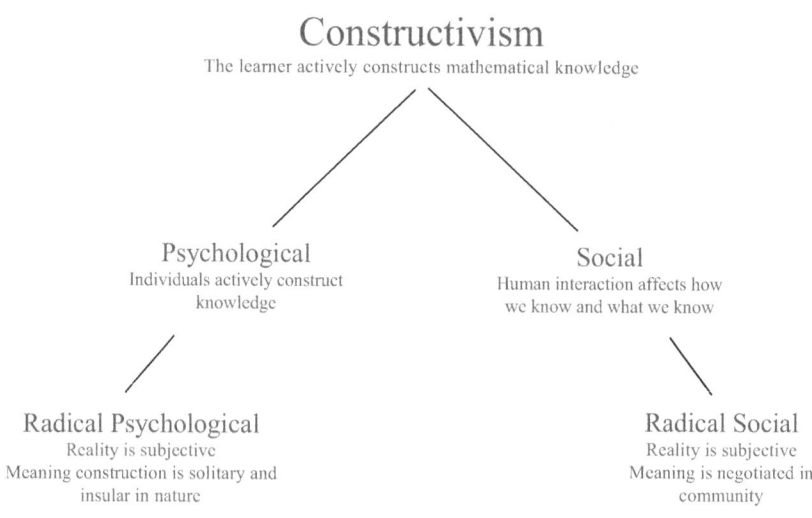

FIGURE 4.1 The Constructivism Family

model of constructivism's major ideological strands and may be used to anticipate some of the relationships and tensions we describe.

Psychological Constructivism

In "Learning as a Constructive Activity," Ernst von Glasersfeld clearly articulates the central tenets of the psychological constructivist thesis: "From an explorer who is condemned to seek 'structural properties' of an inaccessible reality, the experiencing organism now turns into a builder of cognitive structures intended to solve such problems as the organism perceives or conceives" (1983, p. 50). Von Glasersfeld highlights the shift from passive to active approaches to knowing that characterizes constructivist perspectives. Also evident is the psychological constructivist's enthusiasm for knowing as an individual endeavor. We now turn to the thought of one of the most well-known figures in cognitive psychology in order to look more deeply into psychological constructivism and its core tenets.

Piaget's Genetic Epistemology

While Piaget is most famous in the US as the psychologist who developed a stage theory of cognitive development, he also worked in philosophy. Although he shunned the label "philosopher" he did refer to himself as a genetic epistemologist. Kitchener's *Piaget's Theory of Knowledge* explicitly works to establish Piaget as a thinker whose "ideas have larger implications—extending even beyond the field of psychology ..." (Kitchener, 1986, p. 1). At times it seems that Kitchener's project of establishing Piaget as a philosopher is born of a desire to elevate Piaget's status, although in highlighting Piaget the philosopher—or, more specifically, the epistemologist—it is distinctly possible that Kitchener does damage to Piaget's overall contributions. In the preface to the *Dictionary of Genetic Epistemology*, Piaget (as quoted in Kitchener) recognized the constructivist tendency to view psychology and epistemology as practically inseparable:

> genetic epistemology has as its object the examination of the formation of knowledge itself, that is to say of the cognitive relations between the subject and object: thus it bridges the gap between genetic psychology and epistemology in general, which it helps to enrich by considering development.
>
> *(Kitchener, 1986, p. 2)*

As the quotation suggests, Piaget acknowledged the separations between the disciplines of psychology and epistemology, but his own work was located at the union of the two. While Kitchener might be right that Piaget was more than "just a psychologist," it is also true that trying to eliminate the psychology from Piaget's epistemology would render both impotent, as Piagetian epistemology is inherently psychological. As such, both Piaget's theory of cognitive development and his more general epistemology deserve consideration.

Von Glasersfeld credits Piaget with developing a constructivism that offers a unique theory of cognition. He explains that Piaget established that knowledge, contrary to its traditional understanding, does not represent an independent, external reality. Instead, knowledge in the Piagetian sense serves an adaptive function, marking a sharp break with the long history of epistemology in the Western world (von Glasersfeld, 1996, p. 3).[7] Piaget's notion of knowledge as helping humans adapt to their environment has its roots in Darwinian biology. Kitchener (1986) makes the case for a Darwinian reading of Piaget when he claims that the most prominent strand running through Piaget's work is that

> all reality—biological, physical, psychological, sociological, intellectual—is evolving in the direction of progress. This evolutionary direction tends toward increasing equilibrium, and this process of equilibration ... is not due to accident or chance but rather is subject to the same underlying "law" or explanatory principle wherever it is found.
>
> *(Kitchener, 1986, p. 6)*

To understand how Piaget's evolutionary epistemology works, the notions of accommodation and assimilation need to be explained in order to elucidate the concepts of adaptation and equilibrium. Recall that adaptation is the general ability of a person to effectively interact with their environment (Santrock & Yussen, 1992, p. 293). To Piaget, this successful interaction takes place by means of assimilation and accommodation. Assimilation is the way in which individuals incorporate novel experiences into their existing knowledge structure. Accommodation is the adjustment of this structure in light of new, conflicting information that does not fit within existing mental structures. To Piaget, this is primarily a cognitive, rational process, so accommodation entails a mental reordering of sorts (Santrock & Yussen, p. 258). An example of assimilation is when a child comes across a new entity that fits neatly within current classification systems, say a child sees a giraffe for the first time and can comfortably place it within the class of mammals. Accommodation takes place when new information necessitates a reorganization of the very structures, or schemata, in which experience is organized. For example, a child's first experience with atypical mammals such as dolphins, whales, or even more jarringly, the occurrence of a duck-billed platypus might require accommodation, or a mental restructuring, of "mammal."

Piaget's famous stage theory is an outgrowth of his recognition that when individuals cannot fit new knowledge into existing schemes, they seek equilibrium by constructing new levels of cognitive structures.[8] This process takes place in stages, and progress from stage to stage becomes possible when a child has trouble reconciling a growing number of sensory experiences with their existing knowledge structure. Piaget's stages of cognitive development possess "an underlying cognitive-logical structure" (Kitchener, 1986, p. 17) and, accordingly, must unfold in a particular order. Children reach "partial equilibration" with each successive stage until they reach the final stage, formal operations. At this point, logical/abstract reasoning is achieved and the child has reached a state of "true equilibrium" (Kitchener, 1986, p. 20).

Piaget as a Philosopher of Mathematics Education

While Piaget's work is included mostly as a means to show how individual/psychological constructivism in general works, he did do extensive work on the acquisition of mathematical knowledge. This glimpse at his work serves as a concrete example of Piagetian genetic epistemology as well as a particular vision of mathematics. Piaget's understanding of how children acquire[9] elementary mathematical concepts follows his more general theory of cognitive development. Ernest explains this understanding as possessing a dual focus:

> First, on the centrality of children's experience, especially physical interaction with the world. Second, on the unfolding logic of children's thought … These are intertwined, as the child represents its experiences and actions mentally, and transforms these representations by means of the developing sequence of logical operation.
>
> *(Ernest, 1991, p. 184)*

So, to Piaget, children construct their understandings (mathematical and otherwise) through actively engaging with their physical environment, "picking up, putting down, moving, using" (Hersh, 1997, p. 226) and this engagement leads to mental development that follows a consistent, rational course.

The second focus of Piaget's theory of mathematical concept acquisition is on the "unfolding logic" present in children. This is where we can question just how far down the road of constructivism Piaget was willing to go. Ernest believes that in this regard, Piaget did not travel far:

> Piaget was a constructivist, proposing that children create their knowledge of the world. However, he also believes that in the creation and unfolding of their knowledge, children are constrained by absolute conceptual structures, especially those of mathematics and logic. Thus Piaget accepts an absolutist view of knowledge, especially mathematics.
>
> *(Ernest, 1991, p. 185)*

In Chapter 2 of *Genetic Epistemology*, Piaget discusses his stage theory in terms of how children come to possess logical structures capable of understanding increasingly complex phenomena, from recognition of an object's inclusion in a particular class to what Piaget refers to as the "correlative" logic of propositions (p. 39).

Positioning Piagetian constructivism within a post-Kantian schema (to borrow Piaget's own term) highlights the tension between empiricism and rationalism and may help explain why his work is regarded as an indivisible fusion of psychology and epistemology. Constance Kamii, a prolific math education researcher and student of Piaget, affirms that her mentor was keenly interested in "answering epistemological questions scientifically" (2000, p. 4). In his endeavors, he recognized the merits and shortcomings of both empiricism and rationalism, though he did not value them equally:

> While Piaget saw the importance of both sensory information and reason, his sympathy lay on the rationalist side of the fence. His 60 years of research with children was motivated to a large extent by a desire to prove the inadequacy of empiricism.
>
> *(Kamii, 2000, p. 4)*

The inadequacy of empiricism does not suggest its irrelevance, but the inferences and generalizations Piaget drew from decades of experimental psychological research clearly pushed the needle towards rationalism and thus towards the primacy of the individual in forging logico-mathematical knowledge.

Criticism of Piaget's Stage Theory/Epistemology

Piagetian epistemology is evolutionary insofar as it involves adaptation to the environment. It does seem, however, that the virtual inevitability of his system does not capture the broader dynamism suggested by Darwinian evolutionary theory. Evolutionary biologist Stephen Jay Gould, frequently wrote about the misconception of thinking about natural selection as "progressive." In the prologue to *Ever Since Darwin: Reflections in Natural History*, Gould explains that Darwin saw evolution as having no grand purpose and no definitive direction (Gould, 1977, pp. 12–13). Gould frequently warned of the dangers inherent in assuming that evolution is progressive and that it has inevitably led to its crowning achievement, humankind:

> evolution has no direction; it does not lead inevitably to higher things. Organisms become better adapted to their local environments, and *that is all*. The "degeneracy" of a parasite is as perfect as the gait of a gazelle.
>
> *(1977, p. 13, emphasis added)*

Gould's point is that it is inaccurate to think about evolution in terms of a hierarchy in which humans sit at the top. Instead, each species should be considered in terms of its adaptedness to its particular environment.[10] Piaget's stage theory is not an extension of evolutionary biology so much as an application of it to human knowledge acquisition, and as such it need not be a perfect match. Nonetheless, Piaget's failure to recognize Darwin's radical and often hard to digest message that there is no universal pattern (beyond variation and selection) or plan to evolutionary mechanisms leads to a questioning of Piaget's fundamental ideas. If adaptation is a local and idiosyncratic phenomenon, then how useful are Piaget's *universal* stages? In the following chapter, we explore the inadequacies of some forms of evolutionary theory and their possible solutions.

The second critique is more common and particularly interesting in light of the former complaints regarding Piaget's overly universal understanding of mathematical knowledge. To many, Piaget's theory is overly individual. Such critics charge that Piaget's work undervalues the role of the social in knowledge construction (Gergen, 1995, p. 28).[11] Piaget's focus on how each individual personally constructs an understanding of their world—regardless of the fact that, to Piaget, the understandings are similar for all—has steered psychology in a direction that grossly underemphasizes the ways in which we learn cooperatively. This criticism has led to the development of social constructivism, a branch considered later in this chapter. First, we explore the more radical brand of individual/psychological constructivism that has emerged in the wake of Piaget's work.

As far as Piaget's specific ideas regarding mathematics, Ernest, coming from a decidedly radical social constructivist perspective, most likely considered his branding of Piaget as a mathematical absolutist to be an indictment. Piaget's notion that children individually construct their own mathematical knowledge—but that the mathematical knowledge they create is unchanging and predetermined—seems a reasonably defensible position. Perhaps, however, Piaget could more readily defend this point of view if he opted to explain the indubitability of mathematics in terms of the stability of mathematics' representations in our physical world. For example, the incredible consistency of physical models of elementary mathematics seems a reasonable premise upon which to argue for the certainty of mathematics.[12] Instead Piaget posits an understanding of mathematics that is otherworldly, and that might actually have more in common with other absolutist accounts, as any dynamism it suggests is teleological, moving in one preordained direction.

Von Glasersfeld's Constructivism

Ernst von Glasersfeld's radical constructivist position has been a lightning rod of controversy. He positioned his work as a continuation of the Piagetian project of explaining cognition in terms of how individuals construct mental structures based on their interaction with their environments. He tended to focus on the solitary nature of knowledge construction, placing less emphasis on the universal logical structures posited by Piaget. To many, von Glasersfeld's epistemology and theory of cognition render truth hopelessly subjective. The root of such criticism is that his approach suggests the existence of deep divides between individual knowers, as well as between individuals and their environments.

Radicalizing Piaget

Prior to examining von Glasersfeld's brand of radical constructivism, it is instructive to revisit his definition of "run-of-the-mill" constructivism. Recall von Glasersfeld's reasonably non-controversial constructivist credo offered earlier: "The notion that knowledge is the result of a learner's activity rather than of the passive reception of information or instruction" (1991, p. xiv). Von Glasersfeld's particular version is radical and goes further than any standard version of constructivism. He cites Piaget as a profound influence and explains that radical constructivists concern themselves with how our cognition helps "us to cope in the world of our experience, rather than the traditional goal of furnishing an 'objective' representation of the world as it might 'exist' apart from us and our experience" (p. xiv). Von Glasersfeld continues with this extension of Piagetian thought:

> Radical constructivism, I want to emphasize, is a theory of *active knowing*, rather than a traditional theory of knowledge or epistemology. From this standpoint, as Piaget maintained fifty years ago, *knowledge serves to organize experience*, not to depict or represent an experiencer-independent reality.
>
> *(1991, p. xix)*

To the von Glasersfeldian radical constructivist, the lack of a knowable objective reality does lead to sharp divides between an individual, the environment, and other individuals. Meaning construction is solitary and insular in nature. Von Glasersfeld recognizes the loneliness of the individual knower according to his theory of knowing:

> models we construct of other people's ideas and mental operating, regardless of whether we are concerned with their politics or their mathematics, are necessarily hypothetical *because we have no direct access to what goes on inside other people's heads*.
>
> *(1991, p. xvi, emphasis added)*

The way in which individuals go about constructing their knowledge, according to von Glasersfeld, is similar to Piaget's genetic or evolutionary account. Von Glasersfeld points out that, to the extent that radical constructivism redefines knowledge as an "adaptive function," the movement is in at least some debt to the pragmatists and their focus on how the construction of knowledge helps us to live given our environment.[13] In "Learning as a Constructive Activity" von Glasersfeld explains how the radical constructivist understands the relationship between the adaptive function of intelligence and the impossibility of an objective "real world":

> From an explorer who is condemned to seek "structural properties" of an inaccessible reality, the experiencing organism now turns into a builder of

cognitive structures intended to solve such problems as the organism perceives or conceives ... What determines the value of the conceptual structures is their experimental adequacy, their goodness of fit with experience, their viability as a means for the solving of problems, among which is, of course, the never-ending problem of consistent organization that we call understanding. The world we live in, from the vantage point of this new perspective is always and necessarily the world as we conceptualize it.

(1983, pp. 50–51)

The subjectivist reconceptualization—in the sense of the world as existing for each of us only as it does from our particular points of view—of what it means to know has some interesting implications in the realm of mathematics. Von Glasersfeld gives the example of what he calls "the idea of equilateral triangle." He tells a story of a teacher drawing an equilateral triangle on the chalkboard and talking to her class about it. After a description of what it means for a triangle to be equilateral, the students could, at their desks, draw their own equilateral triangles. But, as von Glasersfeld points out, while the teacher and students all may have an understanding of what an equilateral triangle is, none of the figures are actually examples of an equilateral triangle. Of the "real" equilateral triangle, von Glasersfeld says, "Such a structure exists nowhere, except in your heads. Yours, the students', and anyone else's who knowingly uses the term equilateral triangle" (von Glasersfeld, 1995, p. 10).

While it sounds as if von Glasersfeld is charting a course toward a Platonist understanding of mathematics (recall the forms of Plato's "perfect realm"), he explains that this is not so. Plato's non-constructivist explanation posits that the triangle exists in a God-given realm of Forms, while von Glasersfeld's radical constructivist explanation posits a different location:

We can show that straightness and continuity are not abstracted from imperfect sensory impressions, but from the movements of attention in the dynamic construction of images we create in our minds. They are, in fact, what Piaget called operative rather than figurative, or sensory structures, because they are abstracted from operations we carry out ourselves.

(1995, p. 10)

Von Glasersfeld's belief that the triangle exists only "in our heads" gives credence to the claim that "what remains of knowledge for the constructivist extends no further than the edges of the individual mind" (McCarty and Schwandt, 2000, p. 44).[14] Interestingly, von Glasersfeld's characterization of the construction of mathematical knowledge through activity (and, it stands to reason, interaction with an external environment) seems inconsistent with his guiding principles. Whether this admission of a world outside of or beyond the mind is strong enough to quiet his critics remains to be seen. Indeed, von Glasersfeld's theory is viewed as an extreme and marginal way to conceive of thinking and knowing. The impossibility of meaningful connections between individuals suggested by von Glasersfeld's solipsistic account is a threatening premise for most who work in epistemology. For those forwarding absolutist accounts of truth, radical psychological constructivism seems to rule out the possibility of the existence of any objective external basis for knowledge, while to many non-absolutists, radical psychological constructivism refuses to recognize the role of social practices and conventions in the formation of knowledge.

Criticisms of Radical Constructivism

The controversy about radical constructivism described at the outset of this section stemmed from questions regarding the utter subjectivity of truth suggested by von Glasersfeld's model. Perhaps his acknowledgement of the environment's role in human knowledge construction is a blow to the subjectivist critique, as von Glasersfeld's recognition of the contributions of the external seems to suggest that his radical constructivism does not allow individuals to simply "make up" their personal knowledge any which way they please. Ernest echoes this sentiment, explaining how "external reality" influences subjective knowledge construction, serving as "a constraint that ensures the continued viability of the knowledge" (1991, p. 71). Of von Glasersfeld's constructivism, Ernest next points out that it is unable "to account for the possibility of communication and agreement between individuals. For the sole constraint of fitting the external world does not of itself prevent individuals from having wholly different, incompatible even, subjective models of the world" (1991, p. 71).

Von Glasersfeld's radical constructivism stresses the individual and idiosyncratic ways in which people construct knowledge. This focus on mental constructions as adaptations to the environment does seem to offer some externally imposed constraint on the nature of individual constructions. However, this connection to the environment does not seem to offer any reasonable way in which people can understand *each other* and trust that their meanings can be socially shared in any important sense.

The following section will consider social constructivism, a version of constructivism that was developed at least somewhat in response to what some of its proponents feel is the primary weakness of radical constructivism, namely the failure of von Glasersfeld and other radical constructivists to adequately explain the role of social groups in the construction of knowledge. As a lead-in to this treatment of social constructivism, consider von Glasersfeld's salvo fired at the more socially-oriented constructivists:

> This issue has recently been somewhat confused by talk of shared knowledge and shared meanings. Such talk is often misleading because there are strikingly different ways of sharing. If two people share a room, there is one room and both live in it. If they share a bowl of cherries, none of the

cherries is eaten by both persons. This is an important difference ... The conceptual structures that constitute meanings or knowledge are not entities that could be used alternatively by different individuals. They are constructs that each user has to build up for him- or herself.

(von Glasersfeld, 1996, p. 5)

Von Glasersfeld concludes by explaining that the social constructivists (he describes this group as "those who are stressing the social dimensions of language and knowledge") ought to use the term "taken-as-shared"[15] in place of just simply referring to knowledge as shared. This new term, according to von Glasersfeld, calls into attention the "subjective aspect of the situation" by suggesting that one's meanings seem roughly to be the same as others, as opposed to the lock-step agreement suggested by the term "shared meanings." How well social constructivism confronts von Glasersfeld's complaints will be considered in the following section.

Social Constructivism

Social constructivism is a loosely bound group of ways of thinking that reflect a belief in the general tenets of constructivism but a dissatisfaction with what is perceived to be the hyper-individual, even solipsistic tendencies of many forms of psychological constructivism. Kenneth Gergen makes this point forcefully: "The individual self has run its course. The view of the private self ... is no longer viable—not only on conceptual grounds, but in terms of the societal patterns it invites" (1995, p. 181). While shifting focus from the individual to the group, social constructivism does share with other brands of constructivism the same Kantian roots that bring into question the knowability of objective reality. Many social constructivists would even count Piaget among their influences, presumably because Piaget did much to popularize the idea of knowing as a constructive process, although their notion of the social group as the most suitable unit of study directly conflicts with Piagetian individualism. Insofar as Piaget used biological-evolutionary theory to inform psychology, like-minded social constructivists are also indebted to Piaget (Ernest, 1991, pp. 89–91).

Social constructivism differs from its psychological counterpart in its focus on the ways in which human interaction affects how we know and even what there is to know. While the sections that follow present social constructivism largely as a movement that split from individual constructivism within the disciplines immediately related to psychology (and then spread to areas such as education and more recently, to philosophy of mathematics), it needs to be noted that sociologists, anthropologists, critical theorists, and scholars from other disciplines have significantly contributed to this intellectual movement. The clearest example of new disciplinary contributions to social constructivist theory (at least its more radical forms) is most likely found in the emerging field of sociology of knowledge. Perhaps this complex terrain is what Phillips had in mind when he described the constructivist landscape as "nightmarish." Most certainly, it is multifarious and complex.

It is a landscape that includes tensions other than that between the individual/psychological and the social. In addition to constructivism as a theory of knowing, it is a theory of what there is to actually know. In other words, although one could argue that Piaget and von Glasersfeld did confront the nature of knowledge, their theories were primarily about how we come to know and what we believe, not about the nature of what we do know. Sociology of knowledge as a discipline, following Kuhn and others in the "new" philosophy of science, does focus specifically on the objects of knowledge and are worthy of some consideration. Prior to presenting the contributions of the sociologists of knowledge, it will help to look at the intellectual roots of the movement, which are traceable from 19th-century European thinkers Karl Marx and Émile Durkheim through 20th-century American philosopher of science Thomas Kuhn.

Marx and Durkheim: The Roots of Social Constructivism

Marx's socio-political philosophy emerged as a reaction against the widespread idealism of Hegel (Morrison, 1995, p. 43; Coser, 1977, p. 53). In contrast to the Hegelian emphasis on the primacy of ideas as the driving force of history, Marx set out to show the influence of material conditions on human history as well as on contemporary social and political realities. It is the attention to "concrete social structures" that distinguished Marxist thought from the philosophical tradition of the day (Coser, p. 53). In *The German Ideology*, Marx and Engels make explicit the novelty of their thinking: "It has not occurred to any of these philosophers to inquire into their own material surroundings" (Marx & Engels, 1976, p. 6).

Marx developed the notion that although the prevailing beliefs of the day might seem to be self-evident truths, they are in fact, nothing more than expressions of "the class interests of their exponents" (Coser, p. 53). Furthermore, to Marx, an individual's worldview is shaped by the social-economic class to which the individual belongs. As Morrison explains: "Marx therefore reasoned that the ideas individuals have are related to the way they produce and the class relations they form in the system of production" (1995, p. 45).

While the macro-level, socio-political bent of Marx's work might seem out of place in this project, the ideas put forth by Marx have influenced some socially oriented constructivists. If, as Marx claimed, our understandings of reality are shaped by our material circumstances, then it stands to reason that these social arrangements have a significant hand in the construction of our understandings of the world, right down to how we experience school and also how we come to understand mathematics.

To help clarify Marx's influence on some forms of constructivism, we turn to the work of Émile Durkheim, who extended the Marxist emphasis on the social. Durkheim was a French intellectual working in the late 19th and early 20th centuries. His work centered on the establishment of sociology as a freestanding academic discipline. This recognition of the role of the social was certainly influenced by Marxist ideas and also helped to lay the groundwork for further work in sociology, social-psychology, and social constructivism wherever it may be found (Hughes et al., 1995, p. 155). Coser writes of Durkheim's "sociology of knowledge" (p. 139). As will be detailed later, linking sociology and knowledge suggests that there are social origins to our knowledge and that this knowledge is non-absolute and primarily constructed according to social conventions. Whether Coser is taking liberties with Durkheim's ideas—anachronistically applying the late 20th-century term to an earlier, fundamentally different notion—remains to be seen. Regardless, Coser's label does seem to show the debt contemporary social constructivists feel they owe to Durkheim.

In *The Elementary Forms of Religious Life* (1915), Durkheim explores the role of the social in shaping our thought. Religion was the starting point of his inquiry, but he put forth the notion that our social arrangements affect not just our religious thought but other forms as well, including science. From this, Durkheim draws controversial conclusions on how all humans experience their world. Hughes, Martin, and Sharrock (1995) summarize this facet of Durkheim's thought. They explain how Durkheim argued that religious thought was built by looking toward social and not just natural models, and that

> the basic elements of scientific thought are the same as those of religious thought. But religious thought is itself derived from and modeled upon society. Therefore the way in which we think about the natural world, our most basic categories and procedures of scientific thought, are themselves modeled upon, derived from, the structure of society.
>
> *(1995, p. 195)*

The most controversial (and probably least well-substantiated portion) of Durkheim's work is the idea that our basic categories come from our social arrangements. Recall Kant's explanation of time, space, and other categories as basic, a priori components of the mind. Durkheim's claim refutes Kant's idea and instead posits that even these most basic structures are socially determined (Hughes et al., 1995, p. 196). In *Primitive Classification*, Durkheim and Mauss asserted that, although quite historically distant, even logic has social origins (1963, p. 432).[16] Although Durkheim's thesis regarding the social origins of knowledge is not his most enduring overall contribution, the questioning of the certitude of knowledge continued. The remainder of this chapter will explore more contemporary efforts in this regard.

Recent Roots: Thomas Kuhn and The Structure of Scientific Revolutions

Thomas Kuhn's understanding of how science progresses was dramatically shaken by his work in the history of science. The more he studied earlier scientific eras, the more he began to question the traditional conception that contemporary scientific beliefs are the result of a long history of scientists adding to the field-specific knowledge that existed prior to their inquiries. Perhaps more importantly, Kuhn began to question the commonly held assumption that current scientific beliefs were at the end of a historically contingent, progressive movement toward an uncovering of objective truth.

Kuhn recognized that there were more differences than similarities between scientific eras. For example, the traditional conception of scientific progress holds that Newtonian physics is built upon and thus somewhat similar to its theoretical predecessors. However, Kuhn's studies revealed that Aristotle's theories, rather than being the primitive foundations of Newton's thought were actually an entirely different way of looking at the physical world, starting with other fundamental distinctions than Newton's mass, speed, and gravitation.

Briefly stated, in *The Structure of Scientific Revolutions* (1962), Kuhn identified two distinct phases of science: normal science and scientific revolutions. Normal science refers to the time within a scientific discipline when "research is firmly based upon one or more past scientific achievements ... that some particular scientific community acknowledges for a time as supplying the foundation for its further practice" (Kuhn, 1962, p. 10). Normal science, in the Kuhnian sense, is a restrictive time when the direction and interpretation of inquiry is severely limited by the aims, language, rules, and norms governing the discipline during the time. This is a simplified version of what Kuhn refers to as a paradigm. Kuhn acknowledges how doing normal science within the boundaries of a given paradigm can stifle inquiry:

> Closely examined, whether historically or in the contemporary laboratory, that enterprise seems an attempt to force nature into the preformed and relatively inflexible box that the paradigm supplies. No part of the aim of normal science is to call forth new sorts of phenomena; indeed those that will not fit the box are often not seen at all.
>
> *(p. 24)*

Paradigmatic normal science also affects the temperament of those engaging in the work:

> Nor do scientists normally aim to invent new theories, and they are often intolerant of those invented by others.[17] Instead, normal-scientific research is directed to the articulation of those phenomena and theories that the paradigm supplies.
>
> *(p. 24)*

Kuhn's words call attention to just how restrictive paradigms can be. It is clear that according to *The Structure*, scientific breakthroughs, new discoveries, or other forms of innovation do not take place while scientists are engaged in paradigmatic "mop-up" work.[18] This is where Kuhn introduces the crux of his argument, the scientific revolution. The bulk of the book is dedicated to explaining and giving examples of scientific revolution, so our truncated version of it here, while oversimplified, is intended to capture the spirit of Kuhn's thinking while providing a useful sketch of the concept of scientific revolution.

Since work within a paradigm does not encourage novelty, or progress in the traditional sense, Kuhn points to sudden jarring events in the history of science as the means by which new discoveries are made, thereby changing the course of inquiry within a discipline. Although Kuhn provides a reasonably detailed explanation of how work within a paradigm eventually becomes ripe for revolution,[19] his basic premise is that scientific revolutions represent sharp breaks with past ways of thinking about, understanding, and dealing with the phenomena associated with a particular paradigm. While revolutions become more likely when normal scientific work within an existing paradigm fails to account for an increasing amount of phenomena, Kuhn sees the revolutionary turning point as a somewhat mysterious and often unexplainable event:

> the new paradigm, or a sufficient hint to permit later articulation, emerges all at once, sometimes in the middle of the night, in the mind of a man [sic] deeply immersed in crisis. What the nature of that final stage is—how an individual invents a new way of giving order to data now all assembled—must here remain inscrutable and may be permanently so.
>
> *(Kuhn, 1962, pp. 89–90)*

Crisis, in this Kuhnian sense, means that a paradigm's theories no longer seem to adequately explain the phenomena with which they deal. Although it may rattle our modern sensibilities, Ptolemy's geocentric astronomical model served as a perfectly good explanation for and even predictor of planetary motions for a considerable span of time. But when the geocentric Ptolemaic paradigm failed to predict newer and more reliable celestial observations, Copernicus proposed an elegant, heliocentric alternative. Note that Kuhn's version of how important scientific change (in the form of revolutions) comes about is essentially a private and individual act by someone immersed in an extreme and uncertain situation. Furthermore, Kuhn's depiction of the revolutionary scientist "inventing" a new way to explain phenomena is at the core of the common critique of Kuhn's theory as overly relativistic. In summary, Kuhn presents a sharp contrast between periods of normal science and revolution. In doing so, he posits a version of the history of science that is quite different from the traditional linear model of progress. Instead of a convergence on Truth, Kuhn sees science as a pursuit that is characterized by shifting human priorities and temporary points of view.

The Sociology of Knowledge

Sociologists of knowledge, a group of theorists working as radical constructivists at the intersection of sociology and philosophy (Collins, 2000, p. 11; Bloor, 1976, p. ix), have been quite influential in questioning the certainty of knowledge in areas traditionally considered off-limits for such scrutiny. While it is not controversial to assert that literature, political science, and economics are social constructions, the Edinburgh School's "strong program" has continued with the line of inquiry developed by Kuhn, forwarding the controversial position that science and even mathematics should be thought of similarly[20] (Ernest, 1991; 1998). In *Knowledge and Social Imagery*, David Bloor (1976) explains how from the sociological perspective, all knowledge has its origins in social practices and is thus not certain:

> Can the sociology of knowledge investigate and explain the very content and nature of scientific knowledge? Many sociologists believe that it cannot. They say that knowledge as such, as distinct from the circumstances surrounding its production, is beyond their grasp. They voluntarily limit the scope of their own inquiries. I shall argue that this is a betrayal of their disciplinary standpoint. All knowledge, whether it be in the empirical sciences or even in mathematics, should be treated, through and through, as material for investigation.
>
> *(1976, p. 1)*

Clearly, part of the controversy that surrounds the "strong program" is the extreme nature of its proponents' version of constructivism. As Phillips explains, the Edinburgh School contends that "the form that knowledge takes in a discipline can be *fully explained* or *entirely accounted for*, in sociological terms" (Phillips, 2000, p. 8). Understood in this way, the object of a discipline's inquiry, say physics or even mathematics, is completely determined by social forces. As an example, we turn to the world of prime numbers, a mathematical concept stretching back to antiquity. Most of us learned in school that the number 2 is the first, and hence the smallest, prime number. But primality is by no means intrinsic to the number 2. Rather, a number's primality is determined by the current conventional definition of "prime number" as a natural number with exactly two distinct factors, 1 and itself. However, if we were to use an alternate definition, one that was used in the past ("a prime number is a natural number whose only factors are 1 and itself"), it would be reasonable to consider the number 1 to be prime. Thus the primality of 1 "is a matter of definition, so a matter of choice, context and tradition, not a matter of proof. Yet definitions are not made at random; these choices are bound by our usage of mathematics" (Caldwell & Xiong, 2012, p. 2). So "prime" is a dynamic construct with a history of its own, and even mathematical definitions are mitigated by social and practical factors. While the contemporary definition of "prime number" is widely accepted today, it has not always been the prevailing convention, nor have all mathematicians in previous eras been entirely in agreement on the subject.[21] Seen this way, mathematical constructs are indeed contestable and mutable.

While this very brief excursion into the sociology of knowledge is, at best, a cursory description, it is presented as a means to show the changing climate regarding the debate between those forwarding absolutist conceptions of knowledge and those advocating fallibilist accounts of how we know and what there is to know. Additionally, it is important to recognize that there are different strands of influence operating within the various forms of social constructivism existent today.

Vygotskian Constructivism

If Piaget is the modern father of psychological constructivism, then the analogous role in social constructivism is filled by Soviet psychologist Lev Vygotsky. While Vygotsky never forged a socio-psychological theory about mathematics in particular, his work has been influential to many who consider themselves constructivists (Howe & Berv, 2000, p. 30). This influence has been particularly important to those who feel that constructivism overlooks the role of social interaction in favor of a singular focus on the inner workings of solitary knowers (Bruner, 1985, pp. 21–34). Furthermore, as Ernest has noted, there are several individuals attempting to develop Vygotskian approaches to mathematics education.[22]

The Mind in Social Context

Loosely following Fosnot's summary (1996, pp. 18–28),[23] three facets of Vygotsky's work will be considered as a means to explore his contributions to constructivism: his general vision of psychology as following a sociohistorical development, his concept of the zone of proximal development, and the role of symbols in mediating thought. Like Piaget, Vygotsky was not trained strictly as a psychologist. His initial degree was in law (Wertsch, 1985, p. 6). Perhaps it was the fresh perspective that each researcher brought to their work that helped facilitate such innovation. The Russian philosopher and psychologist G.P. Shchedrovotskii argued that it was specifically this outsider's perspective that enabled Vygotsky to "reformulate psychology in the USSR" (Wertsch, 1985, p. 2).

The sociohistorical nature of Vygotsky's psychology is expressed by his "general law of cultural development." This idea, developed in "The Genesis of Higher Mental Foundations" (1981), basically states that behavior and knowledge are cultural and hence social in nature (Cole, 1996, p. 110). Vygotsky explains how individuals come to possess that which is initially cultural: "Any function in children's cultural development … appears on the social plane and then on the psychological plane. First it appears between people as an interpsychological category and then within the individual child as an intrapsychological category" (Vygotsky, 1981, p. 163).

Vygotsky cautions readers to recognize that human interaction ensures that this phenomenon is dynamic and complex: "but it goes without saying that internalization transforms the process itself and changes its structure and function. Social relations or relations among people genetically underlie all higher functions and their relationships" (Vygotsky, 1981, p. 163).

Michael Cole (1996) positions Vygotsky's "general law" as a natural lead-in to the development of the zone of proximal development since for both, the immediate social relations of a child are critical to cognitive (and social) development (p. 111). Vygotsky differentiated between what he called pseudoconcepts and scientific concepts. The former are the naturally constructed or spontaneous concepts that children form as they reflect on their experiences (Fosnot, 1996, p. 18). The latter are the ideas that come from the more highly structured world of the classroom. As Fosnot explains: "[scientific concepts] impose on the child more formal abstractions and more logically defined concepts than those constructed spontaneously. [Vygotsky] perceived them as culturally agreed-upon, more formal concepts" (p. 18). The zone of proximal development was Vygotsky's construction of the place where children moved from pseudo- to scientific concepts. In *Thought and Language*, Vygotsky explains this relationship between the two types of concepts:

> Though scientific and spontaneous concepts develop in reverse directions, the two processes are closely connected. The development of a spontaneous concept must have reached a certain level for the child to be able to absorb a related scientific concept. For example, historical concepts can begin to develop only when the child's everyday concept of the past is sufficiently differentiated—when his own life and the life of those around him can be fitted into the elementary generalization "in the past and now" ... In working its slow way upward, an everyday concept clears a path for the scientific concept and its downward development.
>
> *(Vygotsky, 1986, p. 194)*

Vygotsky continues to explain this zone by adding that each type of concept creates structures that help each grow toward the other. Elsewhere, Vygotsky and commentators make the relationship between the zone of proximal development and his general law of cultural development more explicit. They reconceptualize the zone of proximal development, trading in the notion that it is where spontaneous or pseudoconcepts meet logical or scientific ones for a vision of a place where the psychological/intrapersonal meets the social/interpersonal. More simply put, Daniels explains that the zone of proximal development: "provides the setting in which the social and the individual are brought together" (1996, p. 7).

The third and final facet of Vygotsky's work is his focus on the role of symbols in mediating thought. A brief explanation of Vygotsky's semiotic interactionism will be followed by more concrete examples. Fosnot refers to Vygotsky's semiotic interactionism as "the dialectical interplay between symbol and thought in concept development" (p. 21). She also points out that although those advancing a constructivist understanding of cognition have been interested in and stimulated by Vygotsky's hypotheses regarding the role of symbols in mental development, contemporary research has yielded mixed results as to the veracity and usefulness of the Vygotskian position.

This semiotic interactionism is also tied to the zone of proximal development in that, when the zone is activated, children struggle to use culturally acceptable symbols to communicate their ideas. The symbols that children learn to use are bounded by the particular culture to which they belong. To take this idea a step further, the ways that children can think are bounded by the symbols to which they have access. Hence, culture determines the ways in which its members think.

Vygotsky and A.R. Luria, also a Soviet psychologist, used studies of different groups of people as evidence supporting their theory. A comparison of rural farmers with little to no formal education and more educated groups in the Soviet Union found that not only their speech but also their reasoning skills and perceptual tendencies were tied to the types of lives lived in the particular culture. For example, Luria asked participants to name geometric shapes such as the ones in Figure 4.2.

Rather than use abstract geometric terminology, the farmers named them according to their resemblance to familiar objects such as "plate" and "bracelet." Of particular interest to Luria was whether or not the rural farmers believed the figures were at all alike. He included a typical response in his report: "No, they cannot be alike ... because the first is a coin and the second a moon." By contrast, individuals from educated groups "ignored the 'individual' feature of each figure, isolated the major figure of 'geometrical class,' and made a decision on this basis" (Luria, 1979, p. 65). In other words, more educated subjects perceived the figures to be alike, since both could be categorized as circles. The results of Luria's research contradicted the idea of a universal perceptual apparatus in which individuals generalize and classify figures according to abstract categories. "More educated subjects may classify such stimuli on the basis of a single 'ideal' property," Luria concludes, "but this is not a natural and inevitable achievement of the human mind" (p. 66). Hence, the very way that we think and reason is culturally determined.[24]

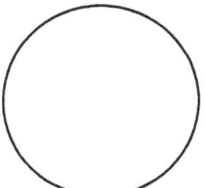 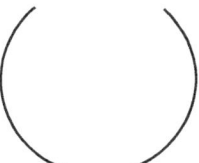

FIGURE 4.2 Images from Luria's research
Source: Luria, 1979, p. 65

Criticisms of Vygotsky's Approach

Some constructivists are disturbed by what they interpret as an absolute truth lurking behind Vygotsky's approach. Keeping in mind that not all constructivists are anti-absolutists in any strong sense of the term, this complaint is only trouble for those claiming that knowledge is highly subjective and relative. Fosnot raises questions on behalf of the worried constructivists: "Is the 'scientific' concept being viewed as 'truth' in the objective sense, and is the teacher's role being perceived as one that facilitates the learner's adoption of it?" (Fosnot, 1996, p. 21). She goes on to question the notion that the child can absorb the formal understandings in the sense that the same meanings can be transmitted from teacher to student. Fosnot sums up these concerns by stating that "These assumptions are not based on the new paradigm (constructivism) but instead are residue of the old. They are still grounded in a theory of learning based on the belief that we hold identical objective meanings" (p. 21). Diana Steele, a modern Vygotskian, uses more current constructivist language as she confronts these very questions. Using Paul Cobb's phrase "taken-as-shared" to describe children's mathematical understandings, she seems to be modifying Vygotsky's ideas given her contemporary context.

A second concern questions the degree to which culture determines everything from behavior to what individuals can actually think. Vygotsky's position seems to leave little room for explanations of variety within a given sociocultural reality. A slightly different twist on the same theme involves questions about how natural or spontaneous concepts are possible at all, given that from infancy, children are exposed to and presumably take in elements of language and hence culture. James Wertsch, a sociocultural anthropologist and Vygotsky scholar, explains this line of criticism because Vygotsky

> neglected [natural development] so completely, he really viewed thinking as the product of social factors alone. For example, S.L. Rubinstein asserted that Vygotsky's account of word meaning elevated the social process involved in speech to the role of the sole creator of thinking.
>
> *(1985, p. 47)*

Although Wertsch goes on to point out that this is a strong reading of Vygotsky and that most would not agree wholeheartedly with Rubinstein, losing the individual to the social is a very real concern with some brands of Vygotsky-inspired social constructivism.

Radical Social Constructivism: The Work of Paul Ernest

To complete the decidedly imperfect analogy started earlier in this chapter, if Vygotsky is the social constructivist counterpart to Piaget's psychological constructivism, then a version of radical social constructivism should be presented as a complement to von Glasersfeld's radical psychological constructivism. To that end, we present the work of Paul Ernest, a social constructivist who does his work in the philosophy of mathematics education.[25] To our knowledge—while certainly not the first to work simultaneously in all three realms: philosophy, mathematics, and education—he is the first to label the field as such. Furthermore, he is a prolific scholar and his application of social constructivism to mathematics education is particularly well developed. In the words of Hersh: "In the young movement to rebuild philosophy of mathematics as part of social reality, Ernest's *Philosophy of Mathematics Education* is one of the most comprehensive and comprehensible" (Hersh, 1997, p. 228). Since 1997, Ernest's *Social Constructivism as a Philosophy of Mathematics* further clarified his research agenda.

Ernest's social constructivist theory is primarily supported by three pillars: the radical psychological constructivist's preoccupation with individual subjective knowledge, the focus on the import of the social and linguistic realms to mathematics, and the quasi-empirical depiction of mathematical knowledge as emerging through acts of human-led recreation. Both of the first two pillars should be recognizable. The earlier sections on psychological constructivism and von Glasersfeld render the notion of subjective knowledge reasonably clear. Likewise, Vygotsky's preoccupation with language as a social tool is evident in Ernest's second pillar. In order to understand Ernest's conception of the role of language in the social enterprise of mathematics it will also be necessary to consider the work of Ludwig Wittgenstein, particularly his conventionalism. The third pillar's quasi-empiricism has not yet been explored in this project, and, accordingly, this facet of Imre Lakatos's work will also be briefly presented.

Ernest's Constructivism: The Influence of Wittgenstein and Lakatos

Briefly stated, the conventionalist conceives of mathematical knowledge and truth as the product of socio-linguistic conventions (Ernest, 1991, p. 30; Hersh, 1997, p. 140). Ernest claims that conventionalism can be absolutist or fallibilist, depending on the form it takes. Ernest is primarily concerned with the alleged conventionalism of Wittgenstein, which was fallibilist in nature (1991, p. 31). We will not attempt to consider the full scope of Wittgenstein's philosophy of mathematics in this section, as it is too broad and deep, not to mention controversial for coverage in this endeavor. Instead, we will consider only Wittgenstein's mathematical conventionalism.

Wittgenstein denies the plausibility of absolutist notions of mathematics. Specifically, he reacts against the Platonist assertion that there are mathematical entities to which our mathematics can and must (if they are to be "true") correspond. He also rejects the disconnected insularity of mathematical formalism. As Klenk explains, to Wittgenstein mathematics merely provides us with forms of inference; in doing mathematics we are simply transforming one expression into another, and whether the transformation is correct is determined not by a correspondence with mathematical objects, but simply by how people actually use these expressions and what they call "correct."

(1976, p. 5)

It is crucial to acknowledge that Ernest's depiction of Wittgenstein is that of a subtle or complex conventionalist. Mathematics, according to this brand of conventionalism, while fallible and contingent, is a practice grounded in a long history of cultural and linguistic practices that shape and even constrain its development. In the words of Wittgenstein: "What is unshakably certain about what is proved? To accept a proposition as unshakably certain—I want to say—means to use it as a grammatical rule: this removes uncertainty from it" (1978, p. 170).

Grammar, to Wittgenstein, suggests that the necessity of mathematics comes from human activity and agreement. Consequently, Wittgenstein's conventionalism recognizes that agreement, convention, and truth have an incredibly complex relationship and that there are many resulting socio-cultural-linguistic constraints on mathematics in addition to mere immediate social agreement. Ernest explains how conventions are

at least partly implicit, since usually no explicit statement of the conventions will be made; rather, participants must infer the conventions from observed behavior and from other's corrections of their own infractions. This might be termed "historical conventions," since it is based on preexisting practices.

(1998, p. 193)

Ernest is borrowing from Wittgenstein's "forms of life," as well as Foucault's idea of a "regime of truth" (Wittgenstein, 1958; Foucault, 1980, 1984). Ernest notes that participating in a "form of life" means following: "the roles and norms and engag(ing) in the expected practices, that is, to observe the *conventions* of the forms of life" (p. 193). Similarly, Foucault's "regime of truth" refers to how a particular worldview (we take this to mean roughly a Kuhnian-type paradigm) affects what is considered to be true. Ernest explains how an idea or theory might seem to be true in the most absolute sense, but that this is only so due to our inability to understand phenomena from any perspective other than our particular worldview: "Such a perspective may seem as well grounded as a foundationalist account of knowledge, but the basis of such truths is the social acceptance and lived nature of the underlying presuppositions" (1998, p. 191).

An example should help illustrate the difference between cruder forms of conventionalism and this Wittgensteinian version. Recall that different definitions of "prime number" lead to different conclusions regarding the primality of 1. Some versions of conventionalism might hold that mathematical facts are whatever is socially agreed upon and that because from the 17th to the 20th century so many mathematicians deemed 1 a prime number, 1 effectively became prime.[26] Wittgenstein and other more complex conventionalists could point to the long history of social practices within the community of mathematicians that would not support such a proclamation. A Wittgensteinian philosopher of mathematics might dwell on how the very ideas of numbers, factors, and divisibility are themselves influenced (if not determined) by socio-cultural and, it follows, linguistic phenomena.[27] Gergen sums this idea up succinctly: "What counts as 'knowledgeable propositions,' then, are dependent on sociohistorical contingencies" (1995, p. 26).

The other major influence on Ernest's work is the quasi-empiricism of Imre Lakatos. Lakatos is generally credited with popularizing quasi-empiricism as a philosophy of mathematics (Ernest, 1991, p. 34; Hersh, 1997, p. 210).[28] He extended Karl Popper's work in philosophy of science, putting forward a conception of mathematics that rejects absolute certainty as a goal and acknowledges the historical, hypothetico-deductive nature of the field. Ernest explains that years prior, Popper had posited that science is based "not on the transmission of truth from true premises to conclusions (the absolutist view), but on the retransmission of falsity from falsified conclusions to hypothetical premises" (Ernest, 1991, p. 35). Quasi-empiricism in mathematics, then, models itself on the ways in which we use the physical world to conjecture about scientific knowledge (Lakatos, 1976; Popper, 1959).

Lakatos coined the term "quasi-empiricism" to show how mathematics and science are similar in method but differ as to content. As Hersh comments:

> His [Lakatos's] own theory is quasi-empiricist, not empiricist *tout court*, because the potential falsifiers or basic statements of mathematics, unlike those of natural science, are not singular spatio-temporal statements (e.g., "the reading of the volt meter was 6.2"). For formalized mathematical theories, he said, the potential falsifiers are informal theories.
>
> *(Hersh, 1997, p. 213)*

So, the Lakatosian view of mathematics is *quasi*-empirical because, while mathematics (like science) does employ tentative verification of its truths, the truths are verified not by physical objects or observations about physical objects but by mathematical ideas.

The result is a version of mathematics that is respectful of its human history of improvement (or at least change) through an unending series of conjectures, counters, and reconfigured conjectures. Lakatos's *Proofs and Refutations* set the stage for including the history of human activity when considering the nature of mathematics. It should be noted that quasi-empiricism is probably more radical to philosophers than it is to mathematicians, as "practicing mathematicians are NOT concerned with philosophical positions *per se* and do what they need to do to get the job done" (K. Parshall, personal communication, May 13, 2002). For the mathematician, empirical considerations are frequently important in helping to "get the job done." In this way, perhaps the most important contribution of Lakatosian quasi-empiricism is the emphasis on the dynamic and imperfect facets of the human endeavor of mathematics, as opposed to any serious light it attempts to shed on the nature of mathematical objects or other ontological issues.

Ernest's Social Constructivism as a Philosophy of Mathematics

Ernest pithily sums up his social constructivist philosophy of mathematics:

> objective knowledge of mathematics exists in and through the social world of human action, interactions and rules, supported by individuals' subjective knowledge of mathematics (and language and social life), which need constant recreation. Thus subjective knowledge recreates objective knowledge, without the latter being reducible to the former.
>
> *(Ernest, 1991, p. 83)*

This condensed description of Ernest's version of social constructivist mathematics does not make clear just what it is that gets constructed. Perhaps the inclusion of a thought experiment proposed by Ernest in the portion of his work preceding the above description will shed light on the products of Ernest's constructivism. As a means to clarify the distinction between private (and, to Ernest subjective) knowledge and public (and it follows, objective) knowledge, Ernest asks the reader to "imagine that all social institutions and personal interactions ceased to exist" (1991, p. 83). Ernest claims that subjective knowledge of mathematics would remain but that objective knowledge would disappear. He explains that

> without social interaction there would be no acquisition of natural language, on which mathematics rests. Without interaction and the negotiation of meanings to ensure a continued fit, individuals' subjective knowledge would begin to develop idiosyncratically, to grow apart, unchecked. The objective knowledge of mathematics, and all the implicit knowledge sustaining it, such as the justificatory canons, would cease to be passed on. Naturally, no new mathematics could be socially accepted either. Thus the death of the social would spell the death of objective mathematics, irrespective of the survival of subjective knowledge.
>
> *(1991, p. 83)*

While the sharp split between public and private may be a questionable philosophical move, his thought experiment does accentuate a constructive element of his social constructivism. Individuals, according to Ernest—in a manner reminiscent of von Glasersfeld's radical constructivism—create or construct their own private understandings of mathematics. Ernest sees the public, social realm as primarily offering a set of constraints and corrections to individual knowledge construction.

Strains of each of the other fallibilist philosophies of mathematics are clearly evident in Ernest's summary of his idea. The radical constructivist's preoccupation with individual subjective knowledge is represented. Likewise, the Vygotskian/ Wittgensteinian focus on the import of the social and linguistic realms to mathematics is evident. Finally, the quasi-empiricist's depiction of mathematical knowledge as emerging through acts of human-led recreation is central to Ernest's social constructivism.

Criticism of Ernest's Radical Social Constructivism

Ernest's mathematical social constructivism is subject to criticisms similar to those leveled at the traditions from which he draws. Quite possibly due to the effects of the polarized nature of philosophy of mathematics discourse, Ernest has opted to draw sharp distinctions between the private–public or inner–outer realms. As a result, Ernest's philosophy, while certainly moving forward constructivist thinking in philosophy of mathematics education, simultaneously possesses some of the problems endemic to both major schools of constructivism. Somewhat counter-intuitively, in upholding some aspects of dualistic thinking, Ernest's thought might also be susceptible to critique for its possession of ideas more often related to the absolutist side of the dualistic divide.

In terms of the independent, unknowable mind suggested by Ernest's brand of constructivism, it seems that Ernest's own critique of von Glasersfeld is in order. Recall that after giving von Glasersfeld credit for developing a psychological theory that did have some ties to an external reality (the environment) and hence seems not to be a form of constructivism that allows one to "just make things up," Ernest warned that

> What the theory [radical psychological constructivism] does not yet do, is to account for the possibility of communication and agreement between individuals. For the sole constraint of fitting the external world does not of itself prevent individuals from having wholly different, incompatible even, subjective models of the world.
>
> *(1991, p. 71)*

Ernest's appropriation of the socio-cultural-linguistic constructivisms of Vygotsky, Wittgenstein, and to a lesser extent, Lakatos is clearly intended to remedy this problem of interpersonal engagement and understanding. Ernest sees social forces as the means by which mathematics gets "out of the head" and into intersubjective agreement. The trouble is, Ernest himself seems to create an unbridgeable chasm between the internal and the external. His thought experiment highlights the idiosyncratic nature of each individual knower's knowledge (in Ernest's parlance it is subjective). So, while the community of mathematicians keeps a fairly steady set of social definitions for mathematics, there is no way to get a sense of whether our personal meanings are even reasonably close to that of our neighbors (taken-as-shared, as Cobb says).

In needing to develop a philosophy of mathematics education that reacts so strongly against absolutism, in some ways Ernest seems to have been unable to shed some of the intellectual baggage of the absolutist tradition. His references to the objective knowledge that is produced through social interactions might immediately raise some constructivist eyebrows and merits scrutiny. Recall that earlier, we explained that Ernest sees conventionalism as a way of thinking about mathematics that could be placed either within the absolutist or constructivist tradition, depending on the specific ways in which a particular form of conventionalism operates. Ernest's conventionalism draws on Wittgenstein and—although Wittgensteinian conventionalism is fallibilist—it seems that Ernest, while calling it "a defensible form of relativism" (1998, p. 247), might still be seeking to find a way to explain the certain Truth that is mathematics. Ernest explains this relativist part of this defensible relativism: "I wish to argue that mathematical knowledge is based on both contingency, due to sociohistorical accident, and deliberate choice by mathematicians, which is elaborated through extensive reasoning. Both contingency and selection are active throughout the long history of mathematics" (1998, p. 248). He goes on to explain how this relativism is not completely unconstrained and hence, defensible: "I also wish to argue that the adoption of certain rules of reasoning and consistency in mathematics means that much of mathematics follows, without further choice or accident, by logical necessity" (p. 248).

Ernest's interest in undermining the absolutist's charge that giving up on the indubitability of mathematics is tantamount to "anything goes" is understandable. However, it seems that Ernest might—possibly for the sake of placating his absolutist critics—be retreating from some of the most interesting and potentially fruitful philosophical terrain. First, by talking about the possibility of objective Truth, Ernest is still adopting the language of his opponents. Perhaps more importantly, by focusing on the adoption of rules and ways of thinking within the community of mathematicians he is ignoring, or at least sidestepping, the question of the ways in which such rules came to be. By defining mathematics as the rules followed by professional mathematicians Ernest is ignoring what he calls the "subjective" knowledge of individuals. The story of how subjective knowledge became "objective" (Ernest's term) needs to be confronted, particularly given Ernest's concern for mathematics education, as it seems that the individual student's struggle to make mathematics meaningful involves much more than the mathematics performed by a group of professionals. A narrow community-of-mathematicians focus overlooks the benefits of including the very real mathematical experiences of many into a definition of mathematics.

Ernest lists four criteria by which to judge the "adequacy" of a philosophy of mathematics:

(i) Mathematical knowledge; its nature, justification, and genesis. (ii) The objects of mathematics; their nature and origins. (iii) The applications of mathematics; its effectiveness in science, technology, and other realms. (iv) Mathematical practice: the activities of mathematicians, both in the present and the past.

(1991, p. 27)

We argue that Ernest's social constructivism is too narrowly focused on the community of mathematicians and, as such, it risks failing according to its own criteria. As to the first pair of criteria, Ernest's social constructivism does offer an account of the nature and justification of mathematics, namely a conventionalist one. However, the genesis and origins of mathematics seem to be forever lost in the long sociocultural history of the practices of groups of mathematicians. If Ernest could expand the scope of what is included in mathematical practice (note that according to his fourth criterion mathematical practice deals exclusively with mathematicians) to include children learning mathematical concepts and everyday users of mathematics then he might be better positioned to be able to offer a glimpse into the "genesis" and "origins" of mathematics.

Constructivism as a Non-Absolutist Alternative

At the end of Chapter 3, we claimed that absolutism fails as "an overarching philosophy of mathematics." As this chapter draws to a close we will add to this judgment the belief that constructivism also does not provide a satisfactory and complete version of the nature of mathematics. While non-absolutist accounts of mathematics are useful insofar as they help to acknowledge the ways in which people and groups actively participate in coming to understand mathematics instead of solving the problems presented by absolutist outlooks, most versions of constructivism tend to create a new set of problems. Frequently, they either react so strongly to what is perceived as an overly rigid explanation of mathematics by countering with wildly subjectivist interpretations that do not seem to adequately count for the remarkable stability of mathematical knowledge or they retain some absolutism by hiding it in descriptions of inevitable structures, constraining language, or in general forms of rationality.

The chapters that follow present some ways of thinking about mathematics that will avoid some of the problems that have plagued philosophy in general and philosophy of mathematics in particular. Attempting to get beyond the inner–outer and individual–social dualisms so prevalent in philosophy of mathematics will not be easy, but it is, we argue, a worthwhile endeavor. While the version of mathematics that is developed will help us to think about mathematics in a more useful way, it is also important to realize that any attempt to find *the* definitive explanation of how mathematics *really is* is bound to create its own share of pathologies, as it is our belief that philosophy in general offers potentially powerful but temporary solutions to problems that are contextually and temporally bound. The evolutionary approach that we develop in Chapter 5 posits the nature of mathematics as a series of evolving concepts and that a particular understanding of mathematics that might have been useful at one time might mostly get in the way at other times. Thus, what follows is presented in the spirit of providing utility given the contemporary contexts in the fields of mathematics, philosophy, and education.

Notes

1 This example was presented by Olive Chapman at the 2017 NCTM Regional Conference in Orlando, Florida.

2 Kitcher makes considerable room for the role of mathematics teachers in his notion of what mathematics is and how we come to know it (1983, p. 91). Hersh also includes mathematics education in his version of the nature of mathematics, claiming "philosophy of mathematics must be compatible with the fact that mathematics can be taught" (1997, p. 237).

3 Arthur Bentley's "The Human Skin: Philosophy's Last Line of Defense" (1954) is an essay criticizing the tendency of philosophers to adopt sharp inner/outer distinctions. We work diligently not to mindlessly do the same. The psychological/social distinction that we have chosen to draw reflects the tendency of the theorists to polarize, not our conception of how things ought to be.

4 This division, while not controversial (see Shotter, 1995, and Richards, 1995, for examples of scholars who use the psychological/social categories), is also not the only way to attempt to organize work that tends to fit the general description "constructivism." For slightly different characterizations, see Matthews (2000).

5 We wish to restate that there are many varieties of constructivism, and by making broad statements about the general claims of the movement we run the risk of not accurately capturing the essence of each variant. However, these general claims will help to explain why opponents of constructivism are disturbed or threatened by what they perceive is undue or damaging subjectivism.

6 Although Descartes sought certainty, it should be noted that in most regards, he never claimed to find it.

7 Von Glasersfeld is right in explaining Piaget's notion of knowledge as an adaptive function; however, he seems mistaken that Piaget is the first to characterize knowledge in such a manner. Late 19th-/early 20th-century American pragmatism is similar in many regards. In Chapters 5 and 6 we present Dewey's version of knowledge as a means to live in and alter an environment. Von Glasersfeld's reading of Piaget is controversial (see the section that follows) and whether such a reading is justified is a legitimate question, but von Glasersfeld's point that Piaget shifted the focus from notions of knowledge as a reflection of reality to knowledge as an adaptive construction seems worthwhile. It also should be noted that von Glasersfeld worked directly with Piaget.

8 For a more comprehensive description of Piaget's theory of cognitive development, any text on educational psychology/child development will be helpful (e.g., Woolfolk, 2018) or from Piaget himself (1969).

9 The word "acquire" appears frequently in the literature related to Piagetian constructivism and, while we go back and forth about whether it connotes sufficient

agency to represent a robust constructivism, there is definitely some agency suggested by the term. This is of interest because it provides a point of entry to consider the post-Kantian attempts at reconciliation of rationalism and empiricism, a difficult reconciliation, to be sure.

10 Kurt Vonnegut's *Galapagos: A Novel* illustrates these strains of evolutionary thinking. It is based on an understanding of the ways in which the human mind is an adaptation that serves well given the environment, and it shows how the idea that human reason is what evolution has been "working" toward is an illusion. Quite disturbingly, *Galapagos* tells the story of a world in which the mechanisms of variation and selection sent the human species in an entirely unanticipated direction.

11 This critique is fiercely rebutted by Kamii (2000): "According to Piaget, the exchange of points of view with others is indispensable for children's development of logic and for scientists' construction of science" (p. 40). She draws heavily on Piaget's 1932 publication, *The Moral Judgment of the Child*, to defend her argument. Kamii also uses the principles of *Moral Judgment* to assert the primacy of children's intellectual autonomy in the math classroom.

12 This line of reasoning bears more than a passing resemblance to J.S. Mill's empirical justification for the certainty of mathematics (a justification that was famously shredded by Frege). Although much maligned, it is an interesting argument that is explored in more detail in the following chapter. (See Mill's *A System of Logic, Ratiocinative and Inductive, Being a Connected View of the Principles of Evidence, and the Methods of Scientific Investigation* (1843/1967).

13 We refer to Piaget's theory of knowing as evolutionary because it parallels biological evolutionary theory in some ways and because Piaget thought of himself as working in a post-Darwinian manner. However, as the criticism of Piaget explained, Piaget's evolutionary theory does not, we believe, go far enough in getting beyond pre-Darwinian (i.e., fixed and absolutist) theories of knowing. In Chapter 5 we develop a more useful version of evolutionary theory in terms of what we know and how we come to know it. The pragmatists, particularly John Dewey's work, will be drawn upon as a particularly well-developed example of evolutionary theory at work in epistemology and, more specifically, with regards to mathematics and its teaching and learning.

14 Note the shift from "head" to "mind." Frankly, we are not certain as to the significance of this change, but "mind" does seem to be potentially more inclusive. Some might even find a way to define "mind" in such a manner as to include physical manipulations.

15 Von Glasersfeld credits Paul Cobb with the coining of the phrase "taken-as-shared" (see Cobb, 1991).

16 Space prohibits further discussion of Durkheim's theory of knowledge. For a brief yet useful explanation, see Morrison, 1995, pp. 200–204.

17 Bernard Barber's "Resistance by Scientists to Scientific Discovery" provides, according to Kuhn, an account of the attitude of scientists on such matters.

18 "Mop-up work" is Kuhn's phrase meaning the work of scientists restricted by a paradigm. See *The Structure*, pp. 23–24.

19 The socio-political ring to this language is purposeful. Kuhn compares scientific and political revolutions in section IX of *The Structure* ("The Nature and Necessity of Scientific Revolutions"). For more on how Kuhn sees paradigmatic normal science giving way to revolution, also see section VI ("Anomaly and the Emergence of Scientific Discoveries").

20 Bloor's work is mostly in the philosophy of science. Paul Ernest, the philosopher of mathematics, has probably forwarded the most well-developed version of social constructivism as a philosophy of mathematics and he draws heavily on Bloor's work (Howell & Bradley, 2001, p. 30). Later in this chapter, a section will be devoted to Ernest's ideas.

21 Whether or not 1 is prime hinges not just on a definition of "prime" but also "number." In fact, Euclid believed that 2 was the first prime number because he did not consider 1 to be a number at all. Rather, it was a unit that served as the building block of numbers, which were "multitudes composed of units." His definition of prime number was a number that is measured "by a unit alone" (Caldwell & Xiong, 2012, p. 3).

22 Just how truly "Vygotskian" these approaches are is subject to some debate, but two such attempts will be given some scrutiny at the end of this section.

23 Our account of Vygotsky also follows, although even more loosely, Harry Daniels' *An Introduction to Vygotsky*, 1996, pp. 1–27.

24 Luria's work can also be used to critique the universality of Piagetian stage theories, particularly Piaget's insistence on abstract rationality as superior to other more adaptive forms of thinking. In another experiment with a group of rural farmers, researchers introduced the statement, "In the far north, where there is snow, all bears are white." This was immediately followed by a premise and question: "Novaya Zemlya is in the far north. What color are the bears there?" Many of Luria's subjects refused to draw any inferences, commenting, for example, that "We always speak only of what we see; we don't talk about what we haven't seen" (Luria, 1979, pp. 77–79).

25 Steffe and Gale (1995) position Kenneth Gergen as von Glasersfeld's "social" counterpart. We have opted to use Ernest because Gergen does not specifically concern himself with mathematics. For interesting comparisons of von Glasersfeld and Gergen, see Steffe and Gale, particularly Shotter's "In Dialogue: Social Constructionism and Radical Constructivism" and Richards's "Construct(ion/iv)ism: Pick One of the Above."

26 It is somewhat disingenuous to say "agreed upon" when the primality of 1 was often in flux within the mathematical community. Famously, Goldbach and Euler reached different conclusions, due in part to the nature of their individual inquiries. For a survey of the history of mathematical primality, see Caldwell and Xiong (2012).

27 The most famous line from Wittgenstein's *Tractatus Logico-Philosophicus* elegantly expresses this dependent relationship: "*The limits of my language* mean the limits of my world" (1922, p. 74).

28 According to Hersh (1994), Lakatos even coined the term quasi-empiricism.

References

Barber, B. (1961). Resistance by scientists to scientific discovery. *Science*, CXXXIV, 596–602.

Bentley, A. (1954). The human skin: Philosophy's last line of defense. In S. Ratnor (Ed.), *Inquiry into inquiries: Essays in social theory* (pp. 195–211). Boston: The Beacon Press.

Bickhard, M. (1998). Constructivisms and relativisms: A shopper's guide. In M. Matthews (Ed.), *Constructivism in science education* (pp. 99–112). Dordrecht: Kluwer.

Bloor, D. (1976). *Knowledge and social imagery*. Boston: Routledge & Kegan Paul.

Bruner, J. (1985). Vygotsky: A historical and conceptual perspective. In J. Wertsch (Ed.), *Culture, communication and cognition* (pp. 21–34). Cambridge: Cambridge University Press.

Caldwell, C.K., & Xiong, Y. (2012). What is the smallest prime? *Journal of Integer Sequences*, 15(9), 1–14.

Chapman, O. (2017, October 18–20). "Making sense of students' alternative mathematical conceptions to inform teaching." Conference session, National Council of Teachers of Mathematics Regional Conference, Orlando, FL, United States. https://www.nctm.org/uploadedFiles/Conferences_and_Professional_Development/Regional_Conferences_and_Expositions/Past_and_Future/Orlando/NCTM_Orlando_Regional_Program_Book.pdf

Cobb, P. (1991). Reconstructing elementary school mathematics. *Focus on Learning Problems in Mathematics*, 13(2), 3–22.

Cole, M. (1996). *Cultural psychology: A once and future discipline*. Cambridge, MA: The Belknap Press of Harvard University Press.

Collins, R. (2000). *The sociology of philosophies: A global theory of intellectual change*. Cambridge, MA: The Belknap Press of Harvard University Press.

Coser, L. (1977). *Masters of sociological thought: Ideas in historical and social context*. Fort Worth, TX: Harcourt Brace Jovanovich.

Daniels, H. (1996). Introduction: Psychology in a social world. In H. Daniels (Ed.), *An introduction to Vygotsky* (pp. 1–27). New York: Routledge.

Descartes, R. (1968). *Discourse on method and the meditations* (F.E. Sutcliffe, Trans.). New York: Penguin Books (Original work published 1637).

Durkheim, E. (1915). *The elementary forms of religious life* (J.W. Swain, Trans.). New York: The Free Press.

Durkheim, E., & Mauss, M. (1963). *Primitive classification*. London: Cohen and West.

Ernest, P. (1991). *The philosophy of mathematics education*. Bristol, PA: The Falmer Press.

Ernest, P. (1998). *Social constructivism as a philosophy of mathematics*. Albany, NY: State University of New York Press.

Fosnot, C. (1996). *Constructivism: Theory, perspectives and practice*. New York: Teachers College Press.

Fosnot, C., & Dolk, M. (2001). *Young mathematicians at work: Constructing number sense, addition and subtraction*. Portsmouth, NH/London: Heinemann.

Foucault, M. (1980). *Power/knowledge*. C. Gordon (Ed.). New York: Pantheon Books.

Foucault, M. (1984). *The Foucault reader*. P. Rabinow (Ed.). London: Penguin Books.

Gergen, K. (1995). Social construction and the educational process. In L. Steffe & J. Gale (Eds.), *Constructivism in education* (pp. 17–39). Hillsdale, NJ: Lawrence Erlbaum Associates.

Gould, S. (1977). *Ever since Darwin: Reflections in natural history*. New York: W. W. Norton.

Hersh, R. (1994). Fresh breezes in the philosophy of mathematics. In P. Ernest (Ed.), *Mathematics, philosophy and education* (pp. 11–20). Bristol, PA: The Falmer Press.

Hersh, R. (1997). *What is mathematics, really?* New York: Oxford University Press.

Howe, K., & Berv, J. (2000) Constructing constructivism, epistemological and pedagogical. In D.C. Phillips (Ed.), *Constructivism in education: Opinions and second opinions on controversial issues*. Chicago: The University of Chicago Press.

Howell, R., & Bradley, W. (2001). *Mathematics in the postmodern age: A Christian perspective*. Grand Rapids, MI: Eerdmans.

Hughes, J., Martin, P., & Sharrock, W. (1995). *Understanding classical sociology: Marx, Weber, Durkheim*. London: Sage.

Kamii, C. (2000). *Young children reinvent mathematics: Implications of Piaget's theory*. New York: Teachers College Press.

Kant, I. (1963). Prolegomena to any future metaphysics which is to be a science. In *Kant* (G. Rabel, Trans.). London: Oxford University Press.

Kitchener, R. (1986). *Piaget's theory of knowledge: Genetic epistemology and scientific reason*. New Haven, CT: Yale University Press.

Kitcher, P. (1983). *The nature of mathematical knowledge*. New York: Oxford University Press.

Klenk, V. (1976). *Wittgenstein's philosophy of mathematics*. The Hague: Martinus Nijhoff.

Kuhn, T.S. (1962). *The structure of scientific revolutions*. Chicago: The University of Chicago Press.

Lakatos, I. (1976). *Proofs and refutations*. Cambridge: Cambridge University Press.

Locke, J. (1979) *An essay concerning human understanding*. P. Nidditch (Ed.). New York: Oxford University Press.

Luria, A. (1979). *The making of mind: A personal account of Soviet psychology*. M. Cole & S. Cole (Eds.). Cambridge, MA: Harvard University Press.

Magee, B. (1997). *Confessions of a philosopher: A personal journey through western philosophy from Plato to Popper*. New York: Modern Library.

Marx, K., & Engels, F. (1976). *The German ideology*. Moscow: Progress Publishers.

Matthews, M. (2000). Appraising constructivism in science and mathematics education. In D.C. Phillips (Ed.), *Constructivism in education: Opinions and second opinions on controversial issues* (pp. 161–192). Chicago: The University of Chicago Press.

McCarty, L., & Schwandt, T. (2000). Seductive illusions: Von Glasersfeld and Gergen on epistemology and education. In D.C. Phillips (Ed.), *Constructivism in education: Opinions and second opinions on controversial issues* (pp. 41–85). Chicago: The University of Chicago Press.

Mill, J.S. (1843/1967). *A system of logic, ratiocinative and inductive, being a connected view of the principles of evidence and the methods of scientific investigation* (8th ed.). London: Longman's, Green and Co.

Morrison, K. (1995). *Marx, Durkheim, Weber: Formations of modern social thought*. London: Sage.

Phillips, D. (1987). *Philosophy, science, and social inquiry*. New York: Pergamon Press.

Phillips, D. (2000). An opinionated account of the constructivist landscape. In D.C. Phillips (Ed.), *Constructivism in education: Opinions and second opinions on controversial issues* (pp. 1–16). Chicago: The University of Chicago Press.

Piaget, J. (1969). The theory of stages (S. Opper, Trans.). In D.R. Green, M.P. Ford, & G. B. Flamer (Eds.), *Measurement and Piaget* (pp. 1–11). New York: McGraw-Hill.

Popper, K. (1959). *The logic of scientific discovery*. London: Hutchinson.

Richards, J. (1995). Construct[ion/iv]ism: Pick one of the above. In L. Steffe & J. Gale (Eds.), *Constructivism in education* (pp. 57–64). Hillsdale, NJ: Lawrence Erlbaum Associates.

Santrock, J., & Yussen, S. (1992). *Child development: An introduction*. Dubuque, IA: Wm. C. Brown.

Shotter, J. (1995). In dialogue: Social constructionism and radical constructivism. In L. Steffe & J. Gale (Eds.), *Constructivism in education* (pp. 41–56). Hillsdale, NJ: Lawrence Erlbaum Associates.

Steffe, L., & Gale, J. (1995). *Constructivism in education*. Hillsdale, NJ: Lawrence Erlbaum Associates.

Von Glasersfeld, E. (1983). Learning as a constructive activity. In J. Bergeron & N. Herscovics (Eds.), *Proceedings of the 5th PME-NA, Montreal 29 September* (2 Vols) (pp. 41–101). Montreal: PME-NA.

Von Glasersfeld, E. (1991). *Radical constructivism in mathematics education*. Norwall, MA: Kluwer Academic.

Von Glasersfeld, E. (1995). A constructivist approach to teaching. In L. Steffe & J. Gale (Eds.), *Constructivism in education* (pp. 3–15). Hillsdale, NJ: Lawrence Erlbaum Associates.

Von Glasersfeld, E. (1996). Introduction: Aspects of constructivism. In C. Fosnot (Ed.), *Constructivism: Theory, perspectives and practice* (pp. 3–7). New York: Teachers College Press.

Vonnegut, K. (1985). *Galapagos: A novel*. New York: Dell.

Vygotsky, L. (1981). The genesis of higher mental functions. In J. Wertcsh (Ed.), *The concept of activity in Soviet psychology* (pp. 144–188). Armonk, NY: M.E. Sharpe.

Vygotsky, L. (1986). *Thought and language*. Cambridge, MA: MIT Press.

Wertsch, J. (1985). *Vygotsky and the social formation of mind*. Cambridge, MA: Harvard University Press.

Wittgenstein, L. (1922). *Tractatus logico-philosophicus*. London: Kegan Paul, Trench, Trubner.

Wittgenstein, L. (1958). *Philosophical investigations*. New York: Macmillan.

Woolfolk, A. (2018). *Educational psychology*. London: Pearson.

PART III

The Alternative: Democratic Mathematics Education

5

EVOLUTIONARY/PRAGMATIC PERSPECTIVES

This book is about the links between mathematics, democracy, and education and why they should matter. We have laid out our arguments in such a way that this chapter marks the farthest point from any explicit treatment of democracy. In it we develop an evolution-inspired frame for thinking about mathematics that is designed to serve specific purposes. First, it helps ease the tension between absolutist and constructivist approaches by linking math neither to a perfect or perfectly removed realm nor to the idiosyncratic beliefs of an individual or group, but to the contexts of the actual lives that humans lead. This focus on evolution also links the realities of our current human existence to possible futures, thus setting the stage for a democratic mathematics education.

At the outset of this book we assert that a main aim of American schooling, at least rhetorically, is the development of individuals capable of and interested in democratic participation and that this aim's conspicuous absence from the mathematics class hurts both school mathematics and the broader democratic education project. Central to our rethinking of mathematics are the ideas of John Dewey, sometimes referred to as the "philosopher of American democracy," and here in Part III—"The Alternative: Democratic Mathematics Education," we present Dewey's thought in stages. Dewey was heavily influenced by Darwin and evolutionary theory and in this chapter, prior to explaining how Dewey thought about mathematics and its teaching, we develop the somewhat controversial argument that evolutionary metaphors can be helpful to understand the origins, nature, and purpose of mathematics. We highlight the evolutionary thinking present in a variety of relevant thinkers, chiefly to prime the reader to be open to this different way of thinking about mathematics. Next we present some of Dewey's ideas about mathematics and mathematics education, drawing primarily on *The Psychology of Number and Its Application to Arithmetic*, an 1895 work co-authored with James

McLellan. We conclude this chapter by introducing a second layer of evolutionary thinking and apply it to mathematics and school mathematics.

In Chapter 6, we fold all of this thinking about mathematics, with its philosophical and epistemological implications in tow, into Dewey's broader democratic theory. In this way, we explicitly relate mathematics education to the broader social and civic aims of school with the intent of making both stronger. These links enrich the philosophy of mathematics that we develop, and they will also provide bridges to the burgeoning critical mathematics education movement. The democratic dimensions of our philosophy of mathematics education will serve as a useful way to confront the contemporary split between those espousing constructivist and critical approaches. We want to be careful not to get too far ahead of ourselves, but making our intentions explicit at this juncture can, we believe, help the reader to proceed with equal intention.

Viewing Chapters 3 and 4 collectively, it is evident that both absolutism and constructivism, while having much to offer, ultimately fail as philosophies of mathematics and even more so as philosophies of mathematics education. Absolutism suggests an understanding of mathematics that captures its unique stability but does not acknowledge its human dimensions. Conversely, constructivism tends to encourage understandings of mathematics that feature human involvement but, in doing so, seem to lose the ability to explain the remarkable stability and universality of mathematical knowledge. Stated in terms of the knowledge–belief dichotomy that was touched on in Chapter 1, absolutists tend to primarily value knowledge (although accounting for how this knowledge is obtained can be a problem for the absolutist). A typical absolutist judges belief to be worthwhile to the extent that it matches some objective mathematical knowledge. Plato's notion that ideas about our changing physical world can only be considered mere belief and that knowledge is only possible when pondering the eternal realm of the Forms is evidence of this sharp distinction between the two. While still maintaining the distinction, constructivist understandings of mathematics tend to elevate the status of belief. Recall that radical constructivists argue against the possibility of the absolutist's objective knowledge.[1] In the end, both absolutism and constructivism fail when attempts are made to explain how knowledge and belief relate to each other in any reasonable way.

Absolutism and constructivism differ in fundamental ways, but we argue that both tend to neglect the influence of empirical considerations on the formation and development of mathematics. Furthermore, both tend not to acknowledge the functional nature of mathematics, instead focusing on its structural elements. The evolutionary approach that we develop in this chapter accounts for both mathematics' stability and its contingent, human-influenced qualities; it also provides for a reasonable relationship between knowledge and belief. In what follows, we present the work of several theorists who, to varying degrees and in different ways, contribute to an overall understanding of the empirical and functional dimensions of mathematics.

Evolution and Mathematics

> Old questions are solved by disappearing, evaporating, while new questions corre-
> sponding to the changed attitude or endeavor and preference take their place.
> Doubtless the greatest dissolvent in contemporary thought of old questions, the
> greatest precipitant of new methods, new intentions, new problems, is the one effected
> by the scientific revolution that found its climax in the Origin of the Species.
>
> *(John Dewey,* The Influence of Darwin on Philosophy, *1920, p. 19)*

Mathematics is a discipline that has been largely untouched by the Darwinian revolution. In essence, it is a last bastion of certainty. Take, for example, the work of Daniel Dennett, a respected philosopher and an avowed evolutionist. In *Darwin's Dangerous Idea* (1995), he develops an argument for adopting the process of evolution by natural selection as an overarching explanatory principle for almost all phenomena. He patiently explains how evolutionary biology works and then develops the thesis that our socio-cultural objects can also be thought of as an evolutionary development. For example, he places cars, libraries, and political freedom all on the tree of life. In this way, Dennett recognizes that just as species could have turned out otherwise (he talks of the vast design space and how small a portion of it is taken up by the "tree of life"), so too could have our socio-cultural artifacts.[2]

Dennett places two things beyond the pale of natural selection: the laws of physics and logic/mathematics. By way of explanation, Dennett cites a thought experiment created by Nicholas Humphrey (1987): if you were required to "consign to oblivion" one of the four wonders that follow, which would it be? The wonders are Newton's *Principia*, Chaucer's *Canterbury Tales*, Mozart's *Don Giovanni*, and Eiffel's Tower. Dennett explains how Humphrey supplies the "correct" answer, claiming that Newton's *Principia* would be the one that had to be erased. His reasoning is that Newton's work was replaceable: "Quite simply, if Newton had not written it, then someone else would—probably within the space of a few years" (p. 139). Humphrey adds emphasis to this point by claiming "The *Principia* was a glorious monument to human intellect, the Eiffel Tower was a relatively minor feat of romantic engineering: yet the fact is that while Eiffel did it *his* way, Newton merely did it God's way" (p. 140).

We do not think that it would be too much of a stretch to assume that if we substituted Russell's *Principia* for Newton's, it would still be selected for very similar reasons.[3] The typical absolutist (and probably most non-absolutists) would argue for the reconstructible and non-contingent nature of mathematics. It is not uncommon to hear talk of mathematics as "God's language."[4] While we agree that mathematical literature is, in some ways, more likely to be able to be recreated than a work in the fine arts, it is not because mathematics corresponds to some external truth. Instead, the reason probably has much to do with the nature of mathematics as a discipline. Those working within the discipline of mathematics operate under reasonably tight systems of social and formal constraint. Mathematics becomes reconstructible in part because other mathematicians work under somewhat similar circumstances with similar aims and, as a result, they will probably be in a position to think along similar lines. Considered in this way, Russell's *Principia* (although a failure in many important ways, as is pointed out in Chapter 3) is just as much a "masterpiece" as the other works and deserves preservation from "consignment to oblivion." The contributions of mathematicians from any particular era are, in a sense, time capsules with valuable information about the communities in which they lived and worked.

This is merely an introduction to this way of thinking about mathematics. Dennett's position is introduced as an example of the widespread notion that, while most things are changing over time, mathematics does not evolve—it just is. On the other hand, Phillip Kitcher makes a case for the worth of an evolutionary approach to philosophy, including thought regarding the nature of mathematics:

> Collectively, mathematical knowledge evolves as successive communities of mathematicians respond to the mathematics they have inherited and to the problems bequeathed to them by natural scientists. The ultimate roots of the tradition lie in relatively primitive manipulations of the environment, carried out by our remote predecessors in India, Babylon, Egypt, and perhaps in sites of which we are ignorant. In the course of the subsequent history, mathematicians have been given a very special role, licensed to devise new languages that relate in ways they find interesting and illuminating to the corpus they have inherited. The demarcation of that role itself represents a discovery about community inquiry, to wit that it is good for other investigations that the role be filled.
>
> *(2003, pp. 410–411)*

Kitcher, a philosopher of science and mathematics, develops an evolutionary account of mathematics that presents a very different version than is the norm. By linking mathematics to changing social and empirical problems in our physical world, he offers a functional account of the origins and development of the discipline. While Dennett seems to marvel at the magnificent structure of mathematics and how it corresponds so perfectly and mysteriously to the natural world, Kitcher (and a few other mavericks) marvel at the way humans have been able to develop mathematics in an effort to improve our lives and solve our problems. In the sections that follow, Kitcher's ideas are presented and augmented with the work of other theorists.

Kitcher's Mathematical Naturalism: A Point of Departure

In "Giving Darwin His Due" (2003), Phillip Kitcher lays out an ambitious agenda for rethinking the knowledge project. It sets the stage for his later application of these ideas to mathematics. It is worth quoting at length as we will, in turn, apply this thinking to philosophy of mathematics education:

> From Descartes to the present, generations of epistemologists have written as though the central problem is to uncover a structure of justification in an individual's beliefs that identifies special warranting relations only among the beliefs themselves or between particular beliefs and the individual's experiences. A far more realistic picture would identify the individual as part of a community, from which much is absorbed, most of it never to be seriously queried, and to view that community as one stage in a historical lineage. Perplexities about particular types of knowledge thus give way to attempts to understand how the pertinent propositions came to be incorporated within the set passed on by the tradition. Further, we can look to Darwin and to the theorists who have succeeded him for clues about how to represent the states of community knowledge at particular times and the transitions among them.
>
> *(Kitcher, p. 412)*

Kitcher's *The Nature of Mathematical Knowledge* offers, he claims early on, a much-needed alternative to mathematical absolutism (or apriorism): "Most of the disputes in philosophy of mathematics conducted in our century represent internal differences of opinion among apriorists … I shall offer a picture of mathematical knowledge which rejects mathematical apriorism" (1983, p. 3). He explains that one reason why absolutism is so prevalent is that philosophers of mathematics have not typically had much else to choose from other than absolutism or some overly simple version of empiricism (1988, p. 294). Kitcher attempts a more careful version of empiricism as a critical component of his overall philosophy of mathematics. Empiricism in this sense refers to the role of our physical or sensory experiences in helping us come to understand mathematics. He goes a bit farther, explaining that his brand of mathematical empiricism has a place for the physical manipulation of objects in actually creating mathematics.

Kitcher's Debt to J.S. Mill's Empiricism

It would be beyond the scope of this project to explore John Stuart Mill's empiricism in great detail. Indeed, we mentioned him only briefly (in an earlier footnote, in fact) when proposing that his description of the origins of mathematics in empirical terms might have helped strengthen a core assumption of Piaget's psychological constructivism. Nevertheless, a bit of background is very much in order here, as Kitcher's empiricism draws on what Kitcher calls a formulation of "Mill's optimal position" (1980, p. 215).

In *A System of Logic*, Mill theorized that mathematics is not to be understood as the study of abstract objects, but instead that it consists of truths that are demonstrated by empirical observations, inductions from experience. As Mill explained, "All numbers must be numbers of something; there are no such thing as numbers in the abstract. Ten must mean ten bodies, or ten sounds, or ten beatings of the pulse" (Mill, 1843/ 1967, p. 167). Hence, the truth of the equivalence relation $2 + 3 = 5$ is intrinsically empirical, as is the non-truth of $2 + 3 = 6$. The mathematics proposed by Mill, one deriving from our experience successfully partitioning a set of five objects into disjoint sets of two and three objects each, is distinctly a posteriori in character.

Mill was not content to relegate arithmetic only to relations between sets of specific instances. While he argued for the empirical origins of mathematics, he also recognized its stability and generalizability:

> But though numbers must be numbers of something, they may be numbers of anything. Propositions, therefore, concerning numbers have the remarkable peculiarity that they are propositions concerning all things whatever; all objects, all existences of every kind, know not our experience. All things possess quantity; consist of parts which can be numbered; and in that character possess all the properties which are called properties of numbers.
>
> *(p. 167)*

Frege's opposition to Mill's empiricism was intense and largely successful, as Frege's lasting influence and the relative obscurity of "Millian" mathematics attests. Briefly stated, Frege's characterization of Mill's theory is that numbers are a quality of groups of objects. Frege used color as an example of a quality that an object either possesses or does not possess and he pointed out that number seems to be a fundamentally different phenomenon. In *Frege, Mill, and the Foundations of Arithmetic*, Gordon Kessler explains Frege's argument:

> Consider a normal property such as *being red*. Given any object, the question whether this object exemplifies this property has a determinate answer. Any object is either red or non-red. This, Frege claims, is a characteristic feature of all properties.
>
> *(Kessler, 1980, p. 66)*

Kessler goes on to explain Frege's account of how numbers are not like color in this regard:

> If I present someone with a complete deck of cards and ask "Does this have the number 1?", there is no determinate answer that can be given. The deck has the number 1 if we are counting decks, the number 4 if we are counting suits, and the number 52 if we are counting cards.
>
> *(p. 66)*

In *Foundations of Arithmetic* (as cited in Kessler), Frege explains that any of these numbers "cannot be said to belong to the pile of playing cards in its own right, but at most to belong to it in view of the way in which we have chosen to regard it" (p. 66).

As for Mill's contention that number comes from arranging objects, Frege stingingly objects, stating that if Mill is right, "what a mercy, then, that not everything in the world is nailed down; for if it were, we should not be able to bring off this separation, and 2 + 1 would not be three!" (1884/1960, p. 9). Frege's critique of Mill effectively kept empiricism out of the mainstream in philosophy of mathematics discussions for more than a century. Kitcher forges a more defensible empiricism to help explain the origins of mathematics. While we touch on this facet of Kitcher's thinking, ultimately we turn to Dewey's oft-overlooked philosophy of mathematics as a means to develop Mill's point of view. Although Dewey was certainly not a simple empiricist, later in this chapter we show that Mill's views on mathematics actually possess latent constructivist-pragmatic tendencies and that Dewey's ideas offer an improvement to Mill's somewhat similar, although embryonic, line of thought.

Kitcher and the Importance of the History of Mathematics to Epistemology

Kitcher states that "most philosophers of mathematics have regarded the history of mathematics as epistemologically irrelevant" (1983, p. 5). The following, taken from Jorge Secada's *Historiography*, is evidence supporting Kitcher's contention:

> If we consider what it would take for a past author to understand contemporary work in different subjects, we arrive at a corresponding range going from dis-cursive conceptual clarification (consider what would be needed to explain, say, non-Euclidean geometries to an ancient Greek mathematician) to substantial exercises of imagination and significant personal transformation (consider what it would take for an archaic Greek poet to come to appreciate James Joyce's *Ulysses* (1922), for example).
>
> *(Secada, 2001, p. 684)*

Secada goes on to draw a sharp line between mathematics and other disciplines:

> The history of certain disciplines—mathematics being the paradigm—seems to be reconstructible synchronically in a way in which the history of other disciplines—literature and art come to mind—is not ... Historical insight is eliminable from the understanding of mathematics in a way in which it is not eliminable from the history of literature.
>
> *(p. 684)*

Kitcher's version of mathematics is very nearly the opposite of Secada's main-stream account. While mathematics might not be quite the same thing as its history, it is certainly a product of it and can be better understood if this history is taken into account (Kitcher, 1988, p. 298). Kitcher's idea can be broken down into two main parts, the origins of mathematics and its development (Kitcher, 1983, p. 96).

The Development and Origins of Mathematics

In *Mathematical Naturalism*, Kitcher starts with a question familiar to readers: how do we come to know mathematics? (1988, p. 297). At this point, he is working primarily within the context of the professional mathematics community and focusing on the established axioms of the discipline, rather than turning to proof as the sole means by which to "know" the veracity of mathematics. Even so, Kitcher turns to formal education:

> In almost all cases, there will be a straightforward answer to the question of how the person learned the axioms. They were displayed on a blackboard or discovered in a book, endorsed by the appropriate authorities, and committed to the learner's memory.
>
> *(Kitcher, 1988, p. 297)*

In short, according to Kitcher, our understanding of mathematics is based in large measure on the mathematics that we were taught in school. Furthermore, in mathematics education, "mathematical knowledge is not built up from the beginning in each generation" (p. 298). Thus, to Kitcher, understandings of the nature of mathematics are determined by the common historically funded version of mathematics that students take in during their school years.

Kitcher's second observation concerns the specific ways in which we come to understand mathematics: "some of this knowledge is acquired with the help of perceptions" (1983, p. 92). Early in our mathematical development we use manipulatives, such as rods and beads, and later we use diagrams to understand more complex mathematics (geometry is one obvious example). Still more advanced mathematics has less in common with "everyday" physical objects but it can, Kitcher argues, be linked to its empirical roots through a succession of transformations.[5]

The idea that an individual can learn mathematics in an isolated and relatively synchronic manner, is to Kitcher, wrongheaded:

> the community supplements primary source (authorities) with *local justifications*, providing the student with ways of looking at mathematical principles which seem to make them obvious. So it comes to *appear* that the mathematician, seated in his study, has an independent, individual means of knowing the basic truths he accepts.
>
> *(1983, p. 93)*

Kitcher enriches the notion of a mathematical community by postulating that the mathematical practices in which the community engages can be conceived of as consisting of five components: a language unique to mathematicians, a set of accepted statements, a set of questions that are taken to be important and not currently settled, a set of reasonings used to justify accepted statements, and a set of views regarding how mathematics is to be done (1988, p. 299). He goes on to summarize his theory of mathematical development:

> I claim that we can regard the history of mathematics as a sequence of changes in mathematical practices, that most of these changes are rational, and that contemporary mathematical practice can be connected with the primitive, empirically grounded practice through a chain of interpractice transitions, all of which are rational.[6]
>
> *(1988, p. 299)*

So Kitcher envisions mathematical development as taking place through a chain of knowers, each bound by the mathematics they learned as well as the conventions particular to communities of mathematicians.

Equally wrongheaded, therefore, is the dismissal of history as a factor in mathematical development and indeed in mathematical understanding. As a contemporary example, in 1995 mathematician Andrew Wiles published a viable proof of Fermat's Last Theorem, a mathematical riddle that had eluded the mathematics community for more than 350 years. While Wiles's proof is unquestionably a "masterpiece of modern mathematics" that hinged on the veracity of a bold conjecture made in 1950 (Singh, 1997, p. 307), it also reflects the accepted practices and structures of rigorous mathematical proofs, practices that were negotiated and even contested over time.[7] Hence, a 17th-century problem required a 20th-century proof not just in content but in method.

In terms of the origins of mathematics, Kitcher is less clear. He certainly offers an account that is at least somewhat in line with Mill's empiricism, albeit a freshly reconsidered version. At times, he sounds very much like a Millian, such as when he explains how this "chain of knowers" began:

> Here I appeal to ordinary perception. Mathematical knowledge arises from rudimentary knowledge acquired by perception. Several millennia ago, our ancestors, probably somewhere in Mesopotamia, set the enterprise in motion by learning through practical experience some elementary truths of arithmetic and geometry.
>
> *(1983, p. 5)*

Kitcher's next move is to link mathematics' origins to its contemporary incarnation: "From these humble beginnings mathematics has flowered into the impressive body of knowledge which we have been fortunate to inherit" (p. 5).

Elsewhere, Kitcher explores both the origins and development of mathematics more deeply, intimating that the originators of mathematics made their mathematical observations in the midst of trying to solve practical problems. This seems a step in the right direction, particularly if Kitcher wants to avoid developing a version of earliest mathematicians that sounds like a cruder version of the solitary thinker sitting in his study—a version that Kitcher himself dismissed as an unacceptable way to conceive of the contemporary mathematician.

This sketch of Kitcher's philosophy of mathematics depicts an unconventional theory that draws on a form of empiricism, history, and community in its effort to provide an alternative to absolutism. However, there are problems with Kitcher's work, including an overly passive account of how we use mathematics, a rigid notion of rationality and the possibility for creativity within his system. Perhaps the most troublesome element of Kitcher's formulation is that in spite of his references to a chain of knowers, Kitcher does not make clear a connection between the empirical origins of mathematics and what he calls its current highly abstract state.

Criticisms of Kitcher's Naturalized Mathematical Empiricism

Although he discusses the practical aims of mathematics (1988, p. 314), throughout his work Kitcher seems to downplay the potential role of the development of mathematics for instrumental purposes. In other words, Kitcher's reference to Mill's empiricism is an adequate explanation of *how* humans recognized the truths of mathematics (they used their senses to see and touch arithmetic). But Kitcher undervalues *why* humans felt compelled to begin the chain of mathematical knowing. The social practicalities[8] and local problems of groups of mathematical knowers, be they professional or nonprofessional, certainly matter in the epistemology of mathematics, particularly within a philosophy of mathematics that attends to historical facets.

Kitcher's inattention to everyday social practicalities is at least partly due to his rigid understanding of rationality. He explains that there are distinct ends of rational inquiry, ends that are "the achieving of truth and the attainment of understanding" (1988, p. 305). He also explains that mathematical developments are "rational insofar as they maximize the chances of attaining the ends of inquiry" (p. 304). The belief that there is a clear and permanent truth toward which mathematical inquiry is headed seems more Hegelian than Darwinian in some important respects.[9] If the development of mathematics had been driven only by events that took place within the insular and abstract world of the community of mathematicians, perhaps this understanding of rationality would be in order. Kitcher sounds as if he views the "attainment of understanding" as a destination. Again, if we think of mathematics as a closed system, then we can also imagine an occasion when inquiry is ended because understanding has finally been reached. But advancing a truly evolutionary theory of the nature of mathematics requires recognition of the influence of dynamic and contingent events from both within the community of mathematicians and from outside of it.

It also requires attention to the phenomenon of change, and Kitcher's explanation of the possibility of novelty in his philosophy of mathematics is insufficient. At the outset of *The Nature of Mathematical Knowledge*, Kitcher addresses the question of how change can come about: "Someone may worry that I have depicted the individual mathematician as subservient to the authority of the community and that I have portrayed the community as dominated by tradition" (1983, p. 10). Kitcher explains that although fledgling mathematicians do enter the field by learning from authorities and that this knowledge does come from the historical development of the discipline, creativity can still have two sources, the first of which "consists in adding to the store of mathematical results without amending the basic framework within which mathematics is done" (1983, p. 11). Kitcher's second form of creativity is more momentous: "Moved by considerations which govern the development of mathematics at all times, a mathematician may modify, even transform, the elements of practice which he inherited from his teachers, introducing new concepts, principles, questions, or methods of reasoning" (p. 10).

Here again, Kitcher's explanation does not take full advantage of what evolutionary theory has to offer a philosophy of mathematics. In biological evolution the randomness of gene combinations and mutations is a source of novelty, and varied environmental circumstances also play a role by serving as agents of selectivity. Thus, new traits emerge because of recombinant genetic material and the environments that partially determine whether the organisms with the new traits will survive and, presumably, pass on the traits. Additionally, novel characteristics are frequently overwhelmed by the "normalness" of a large genetic population. New species come about when novel traits have a large impact on a small, local population and the particular traits are a good fit with the immediate environment. So Darwin found that the isolation of the Galapagos Islands proved a fertile ground for the propagation of unique species because local populations were small and the environment was hospitable to some of the interesting and different characteristics that random mating produced. Likewise with mathematics, random combinations of ideas within local "populations" of mathematicians can lead to ideas with new "characteristics." Whether these new ideas take hold has much to do with the intellectual environment, as changes in the purposes of mathematics or the reasons the expectations the general public has of mathematical inquiry play a role in "selecting" which ideas will be taken seriously.

The story of a chance encounter between physicist Freeman Dyson and mathematician Hugh Montgomery can be illustrative here. One afternoon in 1974, Dyson and Montgomery were among several faculty members to assemble in the tearoom of Princeton's Institute for Advanced Study, an exclusive think tank established in the 1930s (illustrious alumni include Albert Einstein and John von Neumann). After a mutual friend introduced Montgomery to Dyson, the two struck up a conversation about their current pursuits. Montgomery, a young number theorist keenly interested in the famed Riemann hypothesis, mentioned a formula describing the distribution of prime numbers. Dyson recognized the formula instantly, not in relation to prime numbers but rather to the behavior of atomic particles. Their brief exchange altered the course of research on the Riemann hypothesis in that it connected "the deep structure of the basic elements of numbers with that of the basic elements of matter" (Rockmore, 2005, p. 154).

There is an irresistible, though admittedly problematic, comparison between the Galapagos and the Institute for Advanced Study at play here. We want to avoid claiming that a direct and total application of evolutionary theory to philosophy of mathematics is desirable, yet it does seem that thinking about groups of mathematicians working within proximity to each other and working on similar problems might be thought of as local populations.[10] The ideas that each mathematician advances might be thought of as genetic material and what comes out of a local community is bound to be influenced by the particular combination of ideas present. Following through with this metaphor, the work within these local communities and the selective function of the general intellectual environment can be thought of as an explanation of how mathematical change takes place.

Perhaps the most troublesome element of Kitcher's theory concerns the gap he leaves between the pragmatic and empirical origins of mathematics and what mathematics is today. In an effort to explain this gap, Kitcher devotes considerable attention to the transformations that have taken place through a chain of communities of knowers in both *The Nature of Mathematical Knowledge* and in *Mathematical Naturalism*. He argues that mathematics metamorphosed from one kind of activity or form of knowledge into another. But we believe this sharp distinction between types of mathematics is a fundamental error. Granted, mathematics is indeed growing and changing, but we argue instead for an understanding of mathematics that recognizes the *similarities* of the crude empirical understandings of our ancestors and the highly abstract work of contemporary mathematicians, as modern mathematics is not fundamentally different in kind from that which preceded it. Much of the rest of this chapter develops just such a connection.

It may seem that, to this point, our major criticism of Kitcher is that his theory is different from ours. Actually, at the heart of our critique is a belief that Kitcher approaches evolutionary descriptions of facets of mathematics, but that he stops prior to going far enough to gain all of the benefits of this way of thinking. In the end, Kitcher's "mild" evolutionism does not overcome many of the problems of the constructivist outlook, including its typically inadequate explanation for mathematics' unique stability. While our treatment of Kitcher might seem, at times, uncharitable, a thorough critique is necessary because by identifying key shortcomings we begin to see the real potential of an evolutionary perspective. In this regard, we are very much indebted to Kitcher. It is not unlike what mathematicians might say of their historical predecessors, or even of conjectures that do not take us to a particular destination but still point us in a useful direction: "It was a good start. If only they had taken it a step further." Next, we set out to "take it a step further," drawing on the work of several theorists to augment Kitcher's ideas and transform his notion of the nature of mathematics into a full-blooded evolutionary account.

Toward an Evolutionary Framework

Although we argue that Kitcher's theory is not truly evolutionary, it should be noted that it is fairly *revolutionary*. In bringing empiricism back into the conversation through his notion of perceptions as a critical component of the origins of mathematics, Kitcher has also reintroduced psychological explanations of mathematics, as our senses and our psychology are intimately related. Additionally, by linking contemporary practice to the origins of mathematics through a chain of groups of knowers, Kitcher has brought both the history of mathematics and the influence of community into play. Both accomplishments are important.

Kitcher's theory about the origins and development of mathematics serves as our framework for the construction of a more fully evolutionary theory of mathematics. In order to strengthen Kitcher's account, we use Dewey's pragmatic philosophy of mathematics as a means to make the case stronger for the empirical origins of mathematics. Kitcher's reliance on Mill leads to an account of the empirical origins that does not sufficiently recognize the functional role of mathematics in human activity. The result is that Kitcher has trouble explaining how empirical origins have led to the highly abstract nature of contemporary mathematics, and Dewey's functional version of how humans employ empirical objects is one way to connect today's mathematics to its origins, given Kitcher's general framework. Lakatos's historical explanation of the development of mathematics is employed in order to flesh out Kitcher's notion of a "chain of knowers." Situated cognition presents mathematics as a context-dependent and locally-bound activity, countering Kitcher's profile of mathematics that seems restricted to the community of professional mathematicians. Situated cognition can also bridge the gap that Kitcher leaves between the practical origins and the highly abstract and refined mathematics of today. Finally, Toulmin's work on the evolution of concepts is presented as a theoretical explanation of novelty. Toulmin will also serve as the basis for a revisiting of the use of evolutionary theory in philosophy as well as to strengthen the case for its applications, specifically to philosophy of mathematics.

The work of Dewey, Lakatos, the situated theorists, and Toulmin are all intended to add to the effort to "functionalize" Kitcher's semi-evolutionary theory. Mill is used to help explain Dewey's work. Peirce's pragmatism serves as an alternative notion of how mathematics can be thought of as quasi-empirical. The work of Karen Parshall offers an example of a specific use of evolutionary mechanisms as a means to conceive of the development of mathematics. Once the evolutionary approach is fully described, we conclude the chapter by putting these ideas about mathematics in the broader context of Deweyan democratic education.

Dewey's *Psychology of Number*

While John Dewey is regarded as a towering figure in the philosophy of education, he is conspicuously absent in contemporary conversations within the philosophy of mathematics education. This state of affairs is particularly interesting in that, early in his career, Dewey partnered with Canadian educational reformer James McLellan to write a mathematics education book, *The Psychology of Number and Its Applications to Methods of Teaching Arithmetic* (1900).[11] *Psychology of Number* receives little attention from Dewey scholars, philosophers of mathematics, and mathematics educators. For example, in *The Philosophy of Mathematics Education*, Ernest refers to Dewey only in passing. His references paint Dewey with the broad and stereotypical brush of the romantic progressive in the most generic sense (1991, pp. 183–184). This lack of regard/interest among mathematics educators is not uncommon.

In our presentation of Dewey's philosophy of mathematics and mathematics education, we use *Psychology of Number* as the primary text and we draw on other works in order to present a more complete theory. Dewey's philosophy of mathematics shares commonalities with Mill's empiricism and Kitcher's historical conception, although Dewey's version is neither entirely empirical nor historical in orientation. Additionally, Dewey's well-developed explanation of the psychological processes involved in an individual's coming to know mathematics serves to re-inject the human element into a discipline that has frequently worked to explain mathematics in non-psychological terms.

Deweyan Use of "Psychology"

Psychology has often been viewed largely as something to be overcome or ignored in philosophical work, as there is fear that individual mental processes can be a serious impediment to understanding how the world "really is." According to this traditional philosophic conception of psychology there is a sharp line between the mental and the physical. Dewey's philosophy works to mediate between those tendencies that focus disproportionately on either the mental or physical aspects of existence. To combat this artificial, static, and to Dewey, damaging polarization, Dewey employed "psychology" in an unorthodox manner.

Dewey's inclusive and activity-sensitive psychology is at the core of his more general pragmatic beliefs. In "The Postulate of Immediate Empiricism" (1910), he presents an illustration that makes clear why, according to his pragmatic conception of how we know and what there is to know, philosophy and psychology are inextricably linked:

I start and am flustered by a noise heard. Empirically, that noise is fearsome; it really is, not merely phenomenally or subjectively so. That is *what* it is experienced as being. But, when I experience the noise as a *known* thing, I find it to be innocent of harm. It is the tapping of a shade against the window, owing to movements of the wind. The experience has changed; that is, the thing experienced has changed not that an unreality has given place to a reality, nor that some transcendental (unexperienced) Reality has changed, not that truth has changed, but just and only the concrete reality experienced has changed. I now feel ashamed of my fright; and the noise as fearsome is changed to noise as a wind curtain fact, and hence practically indifferent to my welfare. This is a change of experienced existence effected through the medium of cognition.

(Dewey, 1910, p. 230)

So, to Dewey, the world is as it is experienced and vice-versa. His immediate empiricism, or pragmatism, or whatever label we affix to his ideas simultaneously frees and obligates philosophers to employ psychology in understanding experience. This is radical, since many philosophers see psychology as a barrier to logic, obscuring the contents of the logical, a priori realm.

Dewey's reconception of logic and psychology in light of human activity posits psychology and logic as different modes by which we go about figuring out how to live our lives. Psychology is concerned with the mental processes by which we actually think (live), while logic is concerned with the formalization of such "psychological" thinking as a set of norms. Thus conceived, psychology and logic describe different ends of the spectrum of how we know. In *Reconstruction in Philosophy*, Dewey goes as far as discussing the empirical origins of logic:

Logic is a matter of profound human importance precisely because it is empirically founded and experimentally applied ... the problem of logical theory is none other than the problem of the possibility of the development and employment of the intelligent method in inquiries concerned with deliberate reconstruction of experience.

(1920/1967, p. 138)

So to Dewey, logic comes about as fruitful methods of inquiry are recognized, emulated, and eventually their patterns are formalized.

Philosophical Consideration of Number and Mathematics

Dewey conceived of number as transactional in nature—it resides within the process of mathematical activity. In a section titled "Number is a Rational Process, Not a Sense Fact," Dewey explained that we come to use number only after a great deal of rational, abstract thought. The raw sense data with which we work, while rich in gross information regarding the multiplicity of things in nature does not offer any insight with regard to the notion of number: "There are hundreds of leaves on the tree in which the bird builds its nest, but it does not follow that the bird can count" (McLellan & Dewey, 1895, p. 23).

According to Dewey, the process by which number is produced and determined requires both discrimination and generalization. Discrimination involves the recognition that the objects in question consist of distinctly separate units. Consider the number three: a young child faced with three red blocks of identical size and shape first needs to be able to determine that the group of blocks are not one larger unit. Once individual units are discerned, the child next must be able to generalize.

Dewey divided generalization into two sub-processes, the first of which is abstraction. Abstraction requires that the child be able to consider only whatever qualities of an object necessitate its being considered as a part of one group or another. Similarly, they must disregard other distinguishing features, such as the orientation of the red blocks. This abstraction of one or a set of qualities will also eventually include qualities that are not immediately observed by the senses, such as "use or function" (1895, p. 27). Once a child can abstract one or more qualities from objects, the final process in the construction of the elementary notion of number is that of grouping. Grouping requires that the child gather the objects that are deemed similar according to the prior mental (and partially physical) activities of discrimination and abstraction, making a whole. Only through this complex set of processes can crude sense data be considered in terms of number.

Dewey's philosophical explanation of the nature of number requires psychology because, to Dewey, number *is* psychological. Without the processes outlined above, there would only be the ideas of "much" and "many" but not the notions of "how much?" and "how many?" From this psychological explanation of where basic notions of number come from, Dewey next explained how this simple sense of quantity came about as a result of the human need to measure in order to live more efficient and better lives: "Number arises in the process of the exact measurement of a given quantity with a view to ... accurate adjustment of means to end" (p. 42).

Dewey next worked to abolish the distinction that is typically made between counting and measuring. The traditional means of explaining the difference between counting and measuring focuses on how we count to determine how *many* of something there is and how we measure to determine how *much* of something there is. In a sense, this brings us back to Dewey's earlier discussion of discrimination and generalization, that is, whether a phenomenon consists of parts of one whole, or is a related group made up of individual units. To Dewey, the answer was that they may be either, depending upon the context within which these phenomena dwell as well as the needs of the counter/measurer. To illustrate the sameness of counting and measuring Dewey offers a series of concrete examples:

When we count up the number of particular books in a library, we measure the library—find out how much it amounts to as a library; when we count the days of the year, we measure the time value of the year; when we count the children in a class, we measure the class as a whole—it is a large class or a small class, etc. When we count the stamens or pistils, we measure the flower. In short, when we count we measure.[12]

(p. 48)

Dewey next worked to develop his idea that measurement must be undertaken within a larger context. He offered a fairly complex example of how simply measuring a field (determining its area) will not yield a complete picture of its worth. We must know about what the field is capable of producing in order to truly measure it. In other words, we must know how it will affect our lives. Dewey explained that counting/measuring can help us with this problem, as well. For example, the amount of corn, price per bushel, and the cost of tilling the field can all be addressed with measurement. In order to measure in any complex way, the wider context must be accounted for: "All numerical concepts and processes arise in the process of fitting together a number of minor acts in such a way as to constitute a complete and more comprehensive act" (McLellan & Dewey, 1895, p. 57).

Deficiencies in Other Methods

Dewey refers specifically to two methods that he opposes: the "symbols" method and the "things" method. The symbols method teaches number simply as part of a set of symbols. The concept of number that it forwards is essentially an abstract one. According to this method, mathematics primarily involves mastery of the mathematical rules that govern the manipulation of sets of symbols, including the fundamental operations. The second method against which *Psychology of Number* pits itself is the things method. According to the things method, mathematical meaning is derived from physical manipulation of things. Children learn mathematics through exposure to concrete objects. As Dewey explains the things method: "objects of various kinds—beans, shoe-pegs, splints, chairs, blocks—are separated and combined in various ways, and true ideas of number and of numerical operations are supposed necessarily to arise" (p. 60).

Dewey's primary objection was that each method acknowledges only part of mathematics. The symbols method deals only with operations in the mind, with little or no connection to what the operators or operands actually mean. Conversely, the things method deals only with concrete objects and deemphasizes the mental actions that give meaning to numbers and operations. By themselves, each method leaves too much to chance in the sense that we can only hope for connections to be made and for mathematical understanding to emerge. Keeping in mind that, to Dewey, mathematics is the activity by which the mind deals with objects, it becomes easy to see that the two methods combined offer mathematics as mind and mathematics as objects, but the critical component of activity is neglected. As Dewey explained: "it is not the mere perception of things which gives us the idea, but the employing of things in a constructive way" (p. 61). Dewey took what is typically thought of as "out there" and what is often thought of as "within the mind" and recast both within the context of practical experience.

The Two Methods and Rationalism and Empiricism

In his more general philosophy, Dewey frequently depicts his brand of pragmatism as a way to mediate between rationalism and empiricism. His conception of thought as mental *activity in the world* is based upon his more general pragmatic and experimental conceptions of truth, knowledge, and how we understand. In *The Quest for Certainty* (1929), Dewey addresses how neither rationalism nor empiricism captures the dynamic nature of human existence:

> Inquiry proceeds by reflection, by thinking; but *not*, most decidedly, by thinking as conceived in the old tradition, as something cooped up within the "mind." For experimental inquiry or thinking signifies *directed activity*, doing something which varies the conditions under which objects are observed and directly had and by instituting new arrangements among them.
>
> *(Dewey, 1929, p. 123)*

Dewey is clear that the symbols method is either analogous to, heavily influenced by, or actually a branch of rationalism and that the things method possesses a similar relationship to traditional, atomistic empiricism. Dewey uses familiar language to describe the two methods:

> The method of symbols supposes that number arises wholly as a matter of abstract reasoning; the method of objects (things) supposes that it arises from mere observation by the senses—that it is a property of things, an external energy just waiting for a chance to seize upon consciousness.
>
> *(1895, p. 62)*

Once Dewey set up the two methods as analogs to rationalism and empiricism it becomes evident that his philosophy of mathematics is not expressly against everything contained within the two methods. Instead, Dewey wanted to salvage what he could from the two traditions in an effort to advance a philosophy of mathematics and mathematics education that focused on how mathematics is a mental activity involving the objects with which we engage to help us improve our lives. The symbols method can contribute the mental and abstract ingredients, while the things method can help ensure that our mental activities stay tethered to real occurrences in our actual lives. Each facet serves as a phase of the overall activity of thinking, with the things method serving as the inductive phase and the symbols method representing a more deductive phase.

Dewey's Philosophy of Mathematics and Mathematics Education

To Dewey, mathematics is defined by (in fact it actually is) its use. Brute sense data are no more mathematical than random symbol manipulations. The concept of the number three does not reside somewhere within a collection of three apples or beans any more than it does in the symbol "3." "Number is not (psychologically) got *from* things, it is put *into* them" (p. 61). The concept of three is part of the activities in which we engage that require quantification (measuring) as a means to some end.

Dewey used this philosophy of mathematics to develop a philosophy of mathematics education centered on measurement (in its broad Deweyan conception—remember, all counting is measuring and all measuring is counting). Making measurement the vehicle for mathematical explorations ensured, according to Dewey, that number symbols will always be linked to concrete units. Mathematics through measuring encourages an active conception of the discipline. In "John Dewey, E.H. Moore, and the Philosophy of Mathematics Education in the Twentieth Century," historian and Dewey scholar Sidney Ratner writes of the connection between the development of mathematics writ large and the learning of mathematics: "Dewey assumed an analogy between a child's progressive experience with elementary arithmetic and the development of these basic concepts in human history"[13] (1992, p. 105). Furthermore, Dewey justified this course of action for mathematics education by explaining that since mathematics developed from humble origins in measurement over the course of human history, this is a natural way for children to learn the concept of number.

Dewey and Mill

Dewey's philosophy of mathematics sought to offer an alternative to what he saw as the overly empiricist and rationalist options of his day, so it would be easy to conclude that Mill's empirical explanation of mathematical knowledge did not have much in common with Dewey's. While Dewey's version of mathematics emphasized the interplay between empirical objects and our actions, as we hinted earlier, Mill's philosophy of mathematics contained more than just simple empiricism. Mill showed a nascent acknowledgement of the role of human intent in the construction of mathematical knowledge.[14] Mill never fully articulated the pragmatic notion that mathematical knowledge is actually created by the interaction of our activities and the physical world, but he came fairly close: "Two pebbles and one pebble are equal to three pebbles ... affirms that if we put one pebble to two pebbles, those very pebbles are three" (Mill, p. 168). Although not the major point that Mill is attempting to make, here he is clearly including the human act of "putting" the pebbles in his understanding of the nature of the number 3. In *Arithmetic for the Millian*, Kitcher makes a similar point: "We have seen that Mill suggests that apparent references to classes can be parsed away. His idea is to say that objects belong to a class is to assert that *we* regard those objects as associated" (Kitcher, 1980, p. 224). Kitcher makes this point even clearer, stating:

> Thus the root notion in Mill's ontology is that of a collec*ting*, an activity of ours, rather than that of a collec*tion*, an abstract object ... At times, Mill seems to come very close to an explicit proposal of this kind.
>
> *(p. 224)*

The similarities between Kitcher's formulation of Mill's "optimal" position and Dewey's account of mathematics are remarkable. Dewey's general tenet of mathematics as arising from human activity (measuring) certainly is friendly to Mill's account. There are also parallels between Dewey and Mill regarding how children first come to know number. Recall Dewey's description of how abstraction and grouping were required for a child to move from experiencing objects as simply a large or small mass to an understanding of quantity. Dewey's distinction between *how much?* and *how many?* is instructive here. Mill's semi-recognition of the active role of the child in establishing the concept of number is somewhat sympathetic to Dewey's version. Kitcher says of Mill's version of learning arithmetic: "Children come to learn the meanings of 'set,' 'number,' 'addition,' and so forth by engaging in activities of collecting and segregating" (1980, p. 224). To Mill, action is necessary for making meaning of number:

> The fundamental truths of that science all rest on the evidence of sense; they are proved by showing to our eyes and our fingers that any given number of objects—ten balls, for example—may by separation and re-arrangement exhibit to our senses all the different sets of numbers the sums of which is equal to ten. All the improved methods of teaching arithmetic to children proceed on a knowledge of this fact. All who wish to carry the child's mind along with them in learning arithmetic; all who wish to teach numbers, and not mere ciphers—now teach it through the evidence of the senses, in the manner we have described.
>
> *(Mill, 1843/1967, p. 318)*

Mill's distinction between numbers and ciphers is significant, as the word "cipher" even today suggests something secretive and unknowable.[15] In his day, the act of ciphering simply meant doing arithmetic, then a collection of rote computational steps tediously detailed in early American texts such as *Pike's Arithmetic*. Mill's proposal that engaging both the senses and the mind leads to something meaningful stood in stark contrast to pedagogical convention.

Also noteworthy is Kitcher's attempt to explain Mill's overall position regarding the nature of mathematics. Kitcher acknowledges it could be inferred that he is positioning Mill as some type of social constructivist, "one who holds that arithmetic describes constructions which are public rather than nebulous transactions in some mental intuition" (Kitcher, 1980, p. 224). Kitcher explains that he is not calling Mill this sort of constructivist because "it would be more accurate to regard him as concerned less with what we do to the world than with what the world will let us do to it" (p. 224).

Kitcher is on to something here. Mill recognizes that the physical world constrains some of the possibilities of mathematics (at least elementary mathematics). This recognition is an important contribution to an evolutionary philosophy of mathematics, as evolutionary accounts need some explanation of constraints that limit and shape the potential paths development may take. As biological evolution suggests, species that live in one environment tend to develop in ways that differ from species that live in another. For example, animals that live in water are less constrained in terms of an upper limit to their overall weight, as their watery environment makes it so they do not have to support their full body weight, as land-dwelling animals do.[16] It seems that Mill is saying that mathematics, similar to particular species, is not "free" to develop in just any way. The physical world acts as a constraint on the development of mathematics. And whereas Mill's work aids the project of developing an evolutionary understanding of mathematics in this way, Dewey—while similarly acknowledging the role of the physical world—goes further. He adds the critical component of the role that human activity plays in this development.

While the physical world provides some constraints on the potential directions the development of mathematics may take, so too do the ways in which we choose to use mathematics. According to Dewey's pragmatic account, our mathematics is what it is, to a certain extent, because of the ways in which we live our lives. Had we lived our lives differently (by chance, choice, accident, necessity, or other circumstance) it is conceivable that our mathematics might be different. For example, had our ancestors not needed to use geometry to organize their experiences as farmers, the ways in which we systematized spatial relations might be quite different. Just as Kitcher warns about Mill, it would be a mistake to view Dewey as a simple constructivist, as Dewey's focus on function over structure makes clear. In other words, to Dewey, the development of mathematics is driven by the ways in which we use it (i.e., its functions). As was pointed out earlier, a common feature of most different brands of constructivism is their structuralism. According to constructivist accounts, there is some underlying structure that can account for the development of mathematics. Dewey's functionalism suggests a means to consider whether the constructions are good ones—the functional approach tests constructions by acting on them. If the results are satisfactory, then the construction is more than mere belief.

Dewey recognized both the physical and social contributions to the development of mathematics. Kitcher drew on Mill in order to address how humans developed mathematics. In doing so, Kitcher's focus was not on *why* humans developed mathematics. And though Mill's account of the physical world's contributions to mathematics did shed some light on the *how*, Dewey's social version also addresses the *why*. Thus, Kitcher's theory is strengthened by the introduction of Dewey's philosophy of mathematics.

Lakatos and the Psychological/Historical Development of Mathematics

Imre Lakatos, a philosopher of science and mathematics, developed a quasi-empiricist philosophy of mathematics. He compared the discipline of mathematics to a Popperian version of science, putting forward a conception of mathematics that rejects absolute certainty as a goal and acknowledges the ways in which mathematicians develop mathematical knowledge through an unending series of hypotheses and critiques. Recall that Lakatos coined the term "quasi-empiricism" to show how mathematics and science are similar in method but differ as to content. Science uses the physical world as a basis for its experiments while mathematics uses mathematical ideas.

Although the comparison between mathematics and science is interesting and useful, we contend that Lakatos's work is pertinent to this project for another reason. Dewey's work offered a functionalist account of mathematics. His version focused on how lay people develop mathematical understandings. Lakatos, on the other hand, tended to center his study on the ways in which mathematics is undertaken within organized professional (or at least academic) communities of mathematicians where strictly absolutist versions of mathematics are the norm. Lakatosian functionalism is different from Dewey's in that it is more abstract. And whereas Dewey's functionalism considered the ways in which regular people use mathematics to accomplish tasks and understand phenomena in their everyday lives, Lakatosian functionalism offers a fresh perspective on the nature of proof. Rather than thinking of proof as the way mathematicians uncover the immutable Truth that is mathematics, Lakatos conceived of proof more pragmatically. As Kitcher explains in a review of Lakatos's *Proofs and Refutations,* Lakatos's main point is that the role of proofs is not well understood and that this misunderstanding creates problems for mathematical research, and "The mistake is to regard proofs as instruments of justification. Instead we should see them as tools of discovery, to be employed in the development of mathematical concepts and the refinement of mathematical conjectures" (Kitcher, 1977, p. 782).

Dewey thought of mathematics as a set of tools we have developed in order to solve problems in an effort to lead better lives. Lakatos shared an understanding of mathematics that employed the tool metaphor, but his focus differed from Dewey's. Lakatos's work was primarily concerned with the use of mathematics as a tool for inquiry within the discipline of mathematics, be it the professional or school varieties of mathematics.

After a brief consideration of Lakatos's work, as articulated in *Proofs and Refutations*, we reposition Lakatos, particularly regarding what his work adds to Kitcher's account as well as to our own emergent evolutionary approach. Recalling that a major concern of ours is with the ways in which thinking about mathematics can affect mathematics education, it is worth paying particular attention to the rhetorical style of *Proofs*. Its relevance touches on areas as diverse as the educational implications of our work as well as the forging of a connection between the seemingly disparate phenomena of Kitcher's origins and development of mathematics.

Proofs and Refutations

Lakatos described *Proofs and Refutations* as an attempt to cast doubt on the idea that formalist accounts of mathematics are sufficient. He set the stage for his own historically oriented version: "Formalism disconnects the history of mathematics from the philosophy of mathematics, since, according to the formalist concept of mathematics, there is no history of mathematics proper" (1976, p. 2). He went on to point out that formalism fails to accept most of what is typically thought of as mathematics as part of the discipline and that it offers a version of mathematics in which nothing meaningful can be said about its development. In essence, Lakatos was calling for an understanding of mathematics that is not as sterile and ahistorical as the formalist account: "None of the 'creative' periods and hardly any of the 'critical' periods of mathematical theories would be admitted into the formalist heaven, where mathematical theories dwell like the seraphim, purged of all the impurities of earthly uncertainty" (1976, p. 2).

Proofs and Refutations is the fictional story of a discussion set in a mathematics classroom.[17] Lakatos explained that

> the class gets interested in a PROBLEM: is there a relation between the number of vertices V, the number of edges E and the number of faces F of polyhedra—particularly of regular polyhedra—analogous to the trivial relations between the number of vertices and edges of polygons (they are equal)?[18]
>
> *(p. 6)*

The class uses trial and error and comes up with a formula: $V - E + F = 2$.[19] Students surmise (conjecture) that it is likely that this formula will be true for all regular polyhedra. Attempts to falsify the conjecture fail, thus suggesting that a proof will demonstrate the truth of the statement. The bulk of the work is dedicated to what happens once a teacher offers the proof.

Prior to the presentation of this proof, Pupil Sigma suggests that the "proposition seems to be satisfactorily demonstrated" (p. 7), as it held true each of the several times it was tested during the initial trial and error. The teacher opts to present a proof anyway. It involves imagining the polyhedron to be made of rubber, removing one face, stretching the remaining rubber out onto the blackboard and proceeding to use plane geometry techniques to prove the conjecture (involving the addition of a series of diagonals, thus creating triangles and consequently removing each until there is only one left). Upon completion of the demonstration, Pupil Delta remarks that: "You should now call it a *theorem*. There is nothing conjectural about it any more" (p. 8). At this point, the floodgates of doubt open. Pupil Alpha is concerned with the generalizability of this proof: "Are you sure, Sir, that *any polyhedron, after having a face removed, can be stretched flat on the blackboard?* I am dubious of your first step" (p. 8). Before the teacher replies, Pupil Beta questions the certainty of the second step and Pupil Gamma questions the certainty of the idea that dropping the triangles one by one will take away either one edge or two edges and a vertex. Gamma goes on to say: "Are you even sure that *one is left with a single triangle at the end of this process?*" (p. 8). The teacher replies: "Of course I am not sure" (p. 8), and the class embarks on a fascinating inquiry.

What follows is a series of conjectures that are critically considered by the class. Some are rejected outright, sometimes revision is in order. During the course of this discussion Lakatos presented several organizing concepts. Pupil Alpha points out that the attempt at proof turned one conjecture into three and that they are in worse shape than when they started, and Delta asks the teacher: "What does it do then? What do you think a mathematical proof proves?" The teacher's answer introduces the term "quasi-experiment" as a temporary definition of proof, which the teacher declares: "*suggests a decomposition of the original conjecture into subconjectures or of lemmas*, thus *embedding* it in a possibly quite distant body of knowledge" (p. 9). So a lemma is a subconjecture or, in other words, a piece of a proof, or a proof within a proof. As the class continues to think about their potentially problematic lemmas, Lakatos (through the teacher) introduces local and global counterexamples. A local counterexample refutes a particular lemma but does not necessarily cast doubt upon the larger conjecture. A global counterexample is "proof" that the larger conjecture is untenable (p. 11). So false lemmas get exposed by local counterexamples, but the overall proof can be improved by modifying the conjecture. The students eventually come up with what they determine is a global counterexample involving a pair of nested cubes. Rather than accept defeat for his proof, the teacher chooses to make a distinction about what is to be included in his definition of polyhedra. This technique is labeled "monster-barring," as its purpose is to exclude the nested cubes, described as "not a polyhedron at all … a monster,[20] a pathological case, not a counterexample" (p. 14). All the while, the class is reconsidering the ways in which they have chosen to define the terms of their debate, rethinking proof strategies and even stopping to consider what it is they hope to ultimately accomplish with the act of "proving" a conjecture. Lakatos's Pupil Alpha aptly sums up the worth of focusing on the social and nonlinear aspects of proof:

Discovery does not go up or down, but follows a zig-zag path; prodded by coun-terexample, it moves from the naïve conjecture to the premises and then turns back again to delete the naïve conjecture and replace it by the theorem. Naïve conjecture and counterexamples do not appear in the fully fledged deductive structure: the zig-zag of discovery cannot be discerned in the end-product.

(Lakatos, p. 4)

"The zig-zag of discovery" stands in stark contrast to Secada's description of the path from early to modern mathematics as linear and easily reconstituted, without detours, reversals of course, or unplanned stops.

Individual Experiences and the History of the Discipline

Proofs and Refutations would be an impressive achievement even if it simply presented an account of a group of students wrestling with the notion of proof, but there is another dimension to the work. A series of detailed footnotes in *Proofs and Refutations* explains how each development in the class discussion has an analog in the historical development of this particular piece of mathematics. Something quite similar to the initial conjecture was first made by Swiss mathematician Leonhard Euler in 1758 (Lakatos, p. 6) and the footnotes depict a debate that went on well into the 20th century.[21] Each major development in the classroom conversation is paralleled by a historical one. For example, Pupil Sigma initially proclaimed that by virtue of the students' trial and error it ought to be considered a proved theorem. Apparently Euler himself was briefly convinced that he need not do more than offer several examples for which $V - E + F = 2$ (p. 7). The teacher's idea to pretend that a polyhedron was constructed of thin rubber and to lift off a side and stretch what remained onto the blackboard was, according to Lakatos, similar to a proof offered by A.L. Cauchy in 1813 (p. 8).

Lakatos explained how even the questions regarding how to define terms and whether specific instances should be counted as examples of particular phenom-ena have historical analogs. One could interpret his argument to be that the his-tory of mathematics is relevant to its teaching and learning. While surely this was a point of his, we also think that Lakatos had something more radical in mind. As Kitcher says of *Proofs and Refutations*:

What does the example show? I think that Lakatos has demonstrated that there are important issues about mathematical discovery that should not be neglected. The process of mathematical discovery cannot be dismissed (as it so often has been) as a series of "happy guesses."

(1977, p. 196)

In fact, we take a primary contribution of *Proofs and Refutations* to be that however we choose to characterize mathematics, its development ought to be a consideration.

Lakatosian Discovery, Peirce's Pragmatism, and the Evolutionary Perspective

One criticism of *Proofs and Refutations* is that the mathematics it depicts is quite abstract and that the problem and all proposed solutions are largely academic, as they start and end in a mostly theoretical discussion. While it certainly illuminates social dimensions of mathematics that do not frequently see the light of day, the social in this case is strictly within a community of mathematical inquirers. Compare Lakatos's work to Dewey's and it becomes obvious that this is a limited notion of "social." Deweyan mathematics posits a discipline that takes more general human problems as the starting point for mathematical inquiry. In fact, insofar as Dewey's quest was to reunite mathematical theory and everyday practice, it might seem strange to place Lakatos and Dewey so close together. It is our hope that by sharing some of the ideas of C.S. Peirce—a late 19th- and early 20th-century mathematician and a founder of pragmatist philosophy—that the work of Dewey and Lakatos can be rendered complementary if not wholly sympathetic.[22]

In *Logic of Mathematics in Relation to Education*, Peirce acknowledges both the abstract nature of the work within the discipline of mathematics and the idea that frequently such work is driven or at least initiated by more general social problems. At the outset of the article, Peirce establishes this link in a classically pragmatic manner: "A simple way of arriving at a true conception of the mathematician's business is to consider what service it is to which he is called in to render in the course of any scientific or other inquiry" (Peirce, 1898, p. 209). With this statement, Peirce is embedding mathematical activity within a broader context, but he is also establishing that there is a difference between the activities of mathematicians and other inquirers (be they scientific or lay people).

According to Peirce's conception of mathematics, the mathematician's task is to take a "real" problem and "frame another, simpler but quite fictitious problem, which will be within his powers, while at the same time it is sufficiently like the problem set before him to answer, well or ill, as a substitute for it" (p. 210). Peirce goes on to describe the problem once it has been reformulated by the mathematician:

> it is highly abstract. All features that have no bearing upon the relations of the premises to the conclusion are effaced and obliterated. The skeletonization or diagrammatization of the problem serves more purposes than one, but its principle purpose is to strip the significant relations of all disguise.
>
> *(p. 210)*

In this light, perhaps Dewey and Lakatos simply considered different facets of mathematics—Dewey was concerned with problems that are elementary enough to be tackled without taking such a strict disciplinary approach and Lakatos dealt with the problem once it was deemed necessary to involve the mathematician. It is important to note that taking the Lakatosian route toward the abstract does not mean that the inquiry will not have practical ramifications, as advanced branches of mathematics have aided in innumerable developments in the modern world.

There are also similarities in Lakatos's conception of mathematics as a quasi-empirical science and Peirce's explanation of the role of "constructions" in mathematics. Recall that quasi-empirical refers to the way in which mathematics and science have similar methods but that, according to Lakatos, the objects of inquiry differ (the natural world for scientists, mathematical ideas for mathematicians). Peirce explains, referring to Kant, that mathematicians use "constructions" to carry out their inquiries. Constructions, to Peirce, are diagrams, drawings, or models that are created once the initial problem has been purged of its contextual messiness. This abstract model is studied by the mathematician and "new relations are discovered among its parts" (p. 211). Peirce surmises that

> Thus, the necessary reasoning of mathematics is performed by means of *observation and experiment*, and its necessary character is due simply to the circumstance that the subject of this observation and experiment is a diagram of our own creation, the conditions of whose being we know all about.
>
> *(p. 211, emphasis added)*

Peirce invokes observation and experiment to suggest that the constructions are physically operated on by the mathematician quite similarly to the manner in which natural scientists operate on physical phenomena. Manipulating the construction, be it a geometrical diagram or an algebraic equation, can help the mathematician to see relationships that were previously obscured. Furthermore, marks on the paper are, according to Peirce's explanation, empirical objects to be acted upon. Thus, the presentation of Peirce's conceptualization of the mathematician's enterprises serves to more fully develop Lakatos's notion of the quasi-empirical nature of mathematics.

The addition of Lakatos's work to the evolutionary perspective serves to explain mathematics in social terms, as its development is sparked by the interactions of the mathematical community. This is particularly helpful once Peirce's notion of the ties between mathematics and more general pragmatic considerations are brought into play. Just as Dewey's *Psychology of Number* sheds light on the origins of mathematics, Lakatos's *Proofs and Refutations* illuminates some of the ways in which mathematics develops as a discipline. This helps add depth to Kitcher's concept of the "chain of knowers" by offering an example of how inquiry can lead to new understandings. Additionally, Lakatos strengthens the idea that the nature of mathematics and its teaching and learning are connected. Our final chapter looks more closely at this relationship.

The Situated Perspective: Practice, Context, and Mathematics

Some reformers talk of the "social turn" in mathematics education (Lerman, 2000; Boaler, 2000, pp. 1–3). While to some educators this refers to a radical social constructivist perspective, there is a group of individuals working to find a way to bridge the gap between overly psychological and overly social theories of knowledge and cognition. Fosnot refers to a need for a "synthesis ... between those who place more emphasis on the individual cognitive structuring process and those who emphasize the sociocultural effects on learning" (Fosnot, p. 23). Similarly, mathematics educator Jo Boaler explains the origins and nature of this turn:

> knowledge, once regarded as the property of individuals and the bastion of psychologists, may not simply be used in different settings, but emerges as a function of the settings, people, activities, and goals. The implications of this apparently subtle difference are profound ...
>
> *(2000, p. 2)*

John Seely Brown, Allan Collins, and Paul Duguid's "Situated Cognition and the Culture of Learning" sets up an agenda for research on situated theory that fittingly starts with teaching and learning. They boldly claim at the outset of their influential article that the contexts and activities of knowledge creation and use are

> not separable from or ancillary to learning and cognition. Nor is it neutral. Rather, it is an integral part of what is learned. Situations might be said to co-produce knowledge through activity. Learning and cognition, it is now possible to argue, are fundamentally situated.
>
> *(1989, p. 32)*

They go on to implicate education in the artificial separation of thought and action: "The breach between learning and use, which is captured by the folk categories 'know what' and 'know how,' may be the product of the structure and practices of our education system" (Brown, Collins & Duguid, 1989, p. 32). They also articulate a functionalist notion of cognitive activity that is at the core of the various versions of situativity:

> To explore the idea that concepts are both situated and progressively developed through activity, we should abandon any notion that they are abstract, self-contained entities. Instead, it may be more useful to consider conceptual knowledge as, in some ways, similar to a set of tools.
>
> *(p. 33)*

From its origins in anthropologist Jean Lave's context-sensitive, dualism-blurring understanding of thinking to Boaler's applications of this perspective to mathematics education, situativity has been developed as a way to get beyond or at least to de-emphasize the gap between the individual and the social. Although situated theory proponents often are not explicitly employing evolutionary theory, their work seems most at home in this chapter for several reasons. First, as the previous chapter endeavored to show, the most prevalent forms of constructivism are anchored in one of the sides of the individual–social divide. Situated theorists seek to blur this distinction. Furthermore, the various forms of constructivism share a preoccupation with structures. This structuralism often manifests itself in a focus on cognitive structures for psychological constructivists and, not surprisingly, with social structures for social constructivists. Situated theorists' concern with the role of activity in cognition is evidence that, in contrast to the structuralism of most brands of constructivism, situativity forwards a primarily functionalist orientation. This functionalism is, in general spirit and to varying degree, complementary with the evolutionary perspectives of this chapter, particularly as situated theorists attempt to take into account individuals, social groups, and the physical world. In what follows, we very briefly describe how the central idea of the situativity of knowing has affected research, particularly in mathematics education.

Situativity and Anthropology: Lave's Everyday Cognition

Jean Lave's brand of situated theory concerns itself with "everyday" knowing (Rogoff & Lave, 1984). She is a social anthropologist who focuses on recasting learning and related endeavors in terms of social practice. As an anthropologist she noticed that learning and knowing seem to be context-specific activities. This is a radical notion when compared to the more formal theories that are the norm. Her work predates much of the debate in psychology, so it is appropriate to view this as a case where Lave's work in the field, itself an example of situativity, had an impact on theory.

Lave clearly articulates three principles of her situated theory:

1. Learning can be thought of as a facet of all human activity; it's not a separate thing (when you are learning you are actually doing something, not just preparing to do something).
2. Learning is not about knowing knowledge, it is about becoming a certain kind of person and gaining certain kinds of identities in the course of participation in everyday activities in the world.
3. The commonly recognized division between individuals and others is an illusion.[23] (J. Lave, recording, March 10, 1992)

Lave's emphasis on the social points to the use of apprenticeship models as examples of community-level knowing and other real-life situations where mathematics is employed. In *Everyday Cognition* (1984), co-author Barbara Rogoff and Lave explain how a theme of the book is to view thinking as practical activity that is adjusted according to particular situations: "As such, what is regarded as logical problem-solving in academic settings may not fit with problem-solving in everyday situations ..." (p. 7). They go on to state that "in every day actions, thought is in the service of actions. Everyday thinking, in other words, is not illogical and sloppy but instead is sensible and effective in handling the practical problem" (p. 7).[24]

One study where Lave and co-authors Michael Murtaugh and Olivia de la Rocha blended their interest in everyday mathematics and communities of practice involved scrutiny of a very common group: dieting shoppers. In "The Dialectic of Arithmetic in Grocery Shopping" (1984), they study and explain some of the ways in which dieting shoppers solve problems while grocery shopping. The authors observed the shoppers in their "native environments," using anthropological research techniques. Interestingly, the researchers found that the dieters frequently solved their mathematical problems through physical rather than symbolic manipulations. For example, the authors recalled the experience of a new Weight Watchers member when asked by de la Rocha about portion control: "Suppose your allotment of cottage cheese ... is three-quarters of the two-thirds cup the program allows? The problem solver in this example ... after a long pause, suddenly announced that he had 'got it!'" (p. 89). His solution involved filling a measuring cup with two-thirds cup of cottage cheese, dumping it onto a cutting board, dividing it into four portions and serving three of them. Also critical were the settings in which activities took place, as evidenced by the way the dieter with a background in calculus found a suitably contextual solution to his mathematical quandary, and this solution did not involve a set of school-learned, abstract skills.

Situated theory's enthusiasm for multi-disciplinary approaches—e.g., psychology, anthropology, and education—to mathematics and mathematics education seems reasonable, as Lave, Murtaugh, and de la Rocha do present mathematics in a different, less formal light than is generally the case. In a sense, everyday cognition can be viewed as an alternative or even a counterbalance to Ernest's—and perhaps Lakatos's—overly professional and formal version of social constructivism.[25]

Boaler's Situativity and Mathematics Education

In terms of the "math wars," the socially-oriented mathematics educators tend to place their work firmly in the reform camp (see Boaler's introduction to *Multiple Perspectives on Mathematics Teaching and Learning*, 2000) and seem to see it as an important correction to the traditional perspective. Interestingly, they also see their work as a needed corrective to what they describe as the overly individual and internal constructivism that is the inspiration for so many reformers. While those outside of the reform movement probably see little difference between situated cognition and constructivism, Boaler, and others (Boaler, 2000) have taken great pains to point out the differences in their theoretical formulations as well as the differences in practice that each suggests.

Boaler's *Multiple Perspectives on Mathematics Teaching and Learning* is dedicated to exploring the multi- and interdisciplinary possibilities that the situated perspective opens up:

> It is no surprise then that anthropological, sociological, philosophical, political, and other disciplines that have been only minimally represented in mathematics education research in the past are now being employed ... any understanding of learning as participation in different communities of practice will be deeply enriched by the use of multiple perspectives.
>
> *(p. 2)*

Boaler also lays out the differences between the situated and constructivist positions. While it seems true that some constructivist versions of mathematics include a social component, as the individual constructs knowledge based at least partly on what they encounter in their social world, Boaler clears up the difference: "In theoretical terms, constructivism posits a view of learning as the individual mind being influenced by the social world, whereas situated theories propose that learning is a social phenomenon constituted in the world" (p. 5). She goes on to explain this difference in terms of the classroom:

> A student may be given the opportunity to "construct" their own understanding in a mathematics class, by, for example, thinking about a procedure or using it to solve a problem. But if they are not engaging in practices of discussion, procedure adaptation, or modeling, over time, they will be moving along a trajectory of procedure use and they will construct their identities in relation to that.
>
> *(p. 5)*

The implications for math education vis-à-vis situated theory include shifting the focus from the way each student internalizes the subject matter to how students interact with mathematical situations as well as how they interact with other students.

Criticisms of the Situated Approach

The situated perspective does offer mathematics educators a fresh vantage point from which to view their enterprise. Our critique of the movement is lengthier than that of other approaches. This scrutiny is, in a sense, intended as a compliment. The goals of situated theorists (at least in terms of the blurring of what are perceived as problematic distinctions) are quite similar to ours. Thus it is important to carefully analyze how the situated perspective is successful as well as how it is lacking.

Earlier we described situated theory as attempting to offer a synthesis of overly individual and overly social accounts of knowledge and cognition. A different dualism emerges in Rogoff and Lave's explanation that with situated theory, "thought is in the service of action" (p. 7). The potential problems associated with this approach are primarily questions of balance. That is, in abolishing the individual–social distinction it is possible that situated theorists tend to forward an entirely social account of cognition. Lave went so far as to say, "we need to throw out the idea that there are individual learners" (J. Lave, recording, March 10, 1992). Similarly, with regard to the thought–action dichotomy, situated theorists tend to dissolve thought into action rather than seeking to elucidate the interrelationships between the two. Our two primary criticisms focus on situated theorists' tendency toward unbalanced treatments of the individual–social and the thought–action dichotomies. Recognition of dichotomies can be useful, but as discussed in Chapter 2, problems tend to emerge when rigid adherence to the belief that there are only two sides to a given phenomenon restricts our courses of action and ways of thinking. Thinking in this "either–or" manner certainly makes it unlikely, or at least difficult, to consider other options. But we also lose our critical awareness of how the distinctions were drawn by people in a certain time and circumstance, and that those distinctions can be redrawn if the old lines are no longer helpful. We are reminded of a poem by Piet Hein: "Our choicest plans are fallen through, our airiest castles tumbled over, because of lines we neatly drew, and later neatly stumbled over" (Hein, 1966). The point here is that working to blur commonly held distinctions can help to get past some of the problems that came about as a result of drawing the boundaries in such a way in the first place.

Situated cognition proponents run the risk of becoming too rigid in their insistence on designating the social group as the only suitable unit of analysis, thus creating their own essentialized, internal understanding of phenomena (Ortiz, 1999). In this case, the all-important internal–external boundary shifts from being drawn between the individual and everything else to the social group and everything else. Reification of the social rather than the individual carries with it a different but equally problematic set of tendencies. A hard focus on the group can lead to the tendency to assume that communities or individual minds are static, non-complex, and easily understood.

Legal theorist Daniel Ortiz writes of a similar phenomenon within the philosophy and practice of law in "Categorical Community" (1999). The long-running tendency to see individuals as the bounded unit for legal analysis has, according to Ortiz, led to an equally disturbing reactive tendency to see communities as a fixed unit for consideration. Instead, Ortiz argues for "a view of social identity that is more complex and attends to the multiplicity, intersectionality, and instability of the individual's relationship to the community" (p. 769). Likewise, we are concerned about the potential for situated theorists to counter the individual–group dichotomy by reifying various phenomena, depending upon their particular orientations. They run the risk of considering their particular levels of analysis to the exclusion of all others. Earlier we critiqued the tendency of Ernest's social constructivism to view the community of mathematicians as the arbiter of what mathematics is (and also of the radical psychological constructivist's notion of the boundedness of internal mental activity to the exclusion of other knowers). The focus of situated cognition proponents on smaller communities of particular practitioners as the *only* authentic unit of mathematical experience can be similarly troublesome.

The second major criticism centers on what we see as a problematic focus on the informal ways in which we know. Earlier we referred to this second line of criticism as having to do with the balance between thought and action. For the sake of this discussion, we believe the two distinctions can be lumped into one general facet of situated theory.[26] We contend that, at times, situated theorists underappreciate formal bodies of knowledge and institutions. The situated perspective seems to suggest that the formal learning environment of a school is inferior to the more naturally relevant environments offered by individuals out "in the world." While earlier we argued that Ernest's social constructivism seems to acknowledge only the world of formal mathematics and professional mathematicians, viewing education as something that ought to be carried out largely in "the real world" seems an overcorrection. We are reminded of Dewey's discussion at the outset of *Democracy and Education*. He explains that formal education (schools) became necessary when society became sufficiently complex that one could not learn what one needed to be a full participant by simply living in that society. Dewey next explained the dangers inherent in these formal educational settings. He describes how formal education can become

> remote and dead—abstract and bookish, to use the ordinary words of depreciation ... There is the standing danger that the material of formal instruction will be merely the subject matter of the schools, isolated from the subject matter of life experience. The permanent social interests are likely to be lost from view.
>
> *(p. 8)*

While Dewey shares with situated theorists concerns about the abstract nature of the school environment, his educational project sought to infuse the formal institution of school with a heavy dose of relevance. He did not advocate moving education from school to the "real world." Instead his solution seems close to the reverse—working to bring the real world into the school. Sociologist Howard Becker's "A School Is a Lousy Place to Learn Anything In" (1972) offers a sociological look at the relative merits of regular classroom instruction versus apprenticeship models. His conclusion, title notwithstanding, is that while the school setting is far from ideal, apprenticeship models have their own sets of problems. Becker studies apprentice butchers, iron workers, and medical interns, and finds that the informal settings, while rich in the actual practices of the profession or field, frequently do not offer opportunities for apprentices to learn about and actually engage in them. For example, meat markets exist to make money and frequently apprentices are denied meaningful training because they will get in the way of the bottom line. Instead, apprentices are often given jobs such as getting coffee, wrapping the already cut meat, cleaning the tools, etc. He also explains that schools are places where students can experiment relatively safely with what they are learning, as schools are structured in such a way as to make mistakes less damaging. This is clearly not the case with apprenticeship models of education, as improperly cutting a piece of meat can be quite costly to the meat market and failures on the part of the medical intern can result in the loss of life (Becker, 1972, pp. 85–95).

Our point in referring to both Dewey and Becker is that school as a formal institution and mathematics as a formal discipline exist for some very good reasons. The reductions of education to on-the-job training and of mathematics as simply the stuff that people do when they need to "figure" is not a viable solution to the problems associated with overly abstract, formal, and professional understandings of education and mathematics (and mathematics education, for that matter!). To the extent that the situated perspective encourages an over-appreciation of all that is informal, it undermines its own potential worth.

The final criticism is that in confronting the individual–social tension, situated theory tends to underappreciate the role of the environment in creating knowledge. Granted, grounding knowledge in activity implicitly acknowledges environmental influence, as our activity is presumably taking place at least partially as a result of the circumstances we find ourselves in due to the natural world. We acknowledge that a claim of situated theory is that settings matter, and that this ought to go a long way toward easing our concern in this regard. However, "setting" to many situated theorists seems largely socially constructed.[27] In other words, when Lave, Murtaugh, and de La Rocha discuss setting in their study of dieting shoppers, they are talking about the humanly constructed settings of the grocery store or the kitchen and not so much the physical environment in its broadest and most rugged conception. We are looking for situated theorists to also consider the most general claims about the stability of mathematics in our empirical world.

Once again, it is a question of balance—in this case between the role of the social and the external environment—and we argue that situated theorists downplay the role of the environment. This can lead to a version of mathematics that underappreciates its stability.[28] Reacting *against* absolutism can encourage a myopic focus on the psychological and/or social construction of mathematical knowledge. Surely, the stability of mathematics should be explored and woven into any worthwhile philosophy of mathematics.

In spite of the criticisms presented above, the situated approach does have an important role to play in the cultivation of an evolutionary understanding of mathematics. One weakness of Kitcher's philosophy of mathematics is that he adopts a relatively traditional conception of what it means to be rational. In its ideal characterization, the situated approach forwards a notion of rationality that is sensitive to the individual, social contexts, and the physical settings in which thinking/doing is taking place. In short, situated cognition further widens the types of activity that can be considered rational and mathematical.

Evolution and Mathematics: Toulmin's Populational Approach

We have presented the ideas of Kitcher, Dewey, and the situated theorists in an effort to advance a broader understanding of rationality than is the norm. Rationality is not limited to formal logic or abstract thought, but instead is rooted in the contexts of our everyday lives. That is, it involves our ability to adapt to changing circumstances. Additionally, we have argued that mathematics can be thought of as a series of tools that we employ in our rational efforts to deal with the flux of our lives. In *Human Understanding: The Collective Use and Evolution of Concepts*, Stephen Toulmin shares a somewhat similar notion of rationality: "A man [sic] demonstrates his rationality not by a commitment to fixed ideas, stereotyped procedures, or immutable concepts, but by the manner in which, and the occasions on which, he changes those ideas, procedures, and concepts" (1972, p. x).

Presumably, Toulmin is referring to changes that serve particular, well-considered ends and not haphazard changes, or changes simply for the sake of change. His notion of rationality, an evolutionary notion, is quite different from both the static classical model and its relativist opponents. Of the former, Toulmin explains that "the impartial forum of reason was defined as requiring an unchanging system of axioms, or principles; the only question for discussion was, how one should explain the source of those universal principles" (p. 44).[29] Of the latter group, Toulmin notes that although its roots go back to Ancient Greece— he specifies Heraclitus's explanation of how the senses refer "only to particular moments and places" (p. 43)—the momentum for relativism is a more recent phenomenon: "If the eighteenth century places its ultimate reliance on Reason or Nature, and the nineteenth found its intellectual confidence in the providential workings of History, the twentieth century had been plagued by the unsolved problem of Relativity" (p. 49).[30]

So Toulmin sets up his project as offering an evolutionary alternative conception of rationality, given the long-dominant absolutist tradition and its more recently developed relativist rival. We hope that this structure sounds familiar, as a substitution of "philosophy of mathematics" for "rationality" comes quite close to summing up our project. Furthermore, we have argued for stretching the idea of rationality and the nature of mathematics, claiming that mathematics is rational not because it is formal and logical, but *because it helps us solve our problems.*

While Toulmin is suspiciously quiet on the topic of an evolutionary conception of what we typically regard as mathematics, our contention is that a broader functional notion of mathematics fits naturally within the group of phenomena that Toulmin posits will benefit from his "populational approach." As we will soon see, our ability to successfully include mathematics in this group hinges on whether we successfully make the case that mathematics is a "historical entity" and Kitcher, Lakatos, and Dewey certainly seem to help in this regard. Additionally, Lakatos and Peirce's depictions of mathematics as having much in common with the natural sciences can bring us another step closer to thinking about mathematical knowledge in populational terms. Quasi-empiricism renders mathematics less non-contingent, a priori, and logically deductive, and helps to advance the idea that there are mathematical ideas that might be somewhat similar, but that vary slightly as to specific characteristics.

The Populational Approach

Toulmin's populational approach is modeled on Darwin's explanation of change in species. Toulmin focused on Darwin's conception of each species as consisting of "a population which modifie(s) itself in response to the environment" (Devettere, 1973, p. 449). In order for this modification to be possible, there must be considerable intra-species variation, as without this variation the process of selection according to which traits fit best with the given environment would have no impact. For example, if all giraffes were roughly the same height and possessed similar necks, then natural selection would have had no means to encourage the propagation of those who could live best in an environment in which the primary food source (relatively free of competition from other species) was located high in trees.

Toulmin's use of the Darwinian mechanisms of change through variation and selection is his way to find a plausible explanation of scientific novelty and change. Toulmin's work was at least partially a response to Kuhn's *Structure of Scientific Revolutions*. Recall from the preceding chapter that Kuhn's *Structure* proposed that the way that scientific change takes place is through sudden and somewhat unexplainable, unorthodox revolutionary breakthroughs. Kuhn argued that most of the work that gets done in scientific disciplines takes place within the existing paradigm and that a sudden revolutionary change in direction is required for any meaningful change to take place. Implicit in his theory is the idea that each paradigm is merely different from, and not necessarily an improvement over, its predecessors. In fact, to Kuhn, the paradigms are largely incommensurate or incomparable. Toulmin's primary criticism of Kuhn's explanation is that declaring that scientific changes take place by revolution does not really shed any light on the nature of the change. Toulmin's adoption of Darwin's ideas is intended to help in this regard:

> The phrases "conceptual populations" and "selective perpetuation" of variants imply that our analysis should be an "evolutionary" one, not just in the broad sense of being non-revolutionary, but in a quite precise and strict sense

of the term. For the nature of populational change, regarded as a general type of historical process, is already well understood on one special case: viz., that of organic species.

(p. 134)

Toulmin argues that the historical entities are the various disciplines, with species serving as biological analogs. At any given time the disciplines are made up of populations of concepts that vary in a manner somewhat similar to the ways in which biological populations vary (Thackray, 1974, p. 80). Toulmin further supplies a definition of discipline, declaring that a collection of people working in a common area becomes a discipline when their

> shared commitment to a sufficiently agreed set of ideals leads to the develop-
> ment of an isolable and self-defining repertory of procedures; and where those
> procedures are open to further modification, so as to deal with problems arising
> from the incomplete fulfillment of those disciplinary ideals.
>
> *(p. 359)*

Toulmin's definition of a discipline allows for enough coherence that the ideas and activities taking place "within" it are identifiable in contrast to others but not so constraining that there is no variety, thus allowing reasonably diverse "populations" of concepts to exist at any given time. Presumably, the aims of the discipline as a whole drive the selection process. In his review of *Human Understanding*, David Hull illustrates this characteristic balance of coherence and variation by way of an example: "The continuity [of Mendelian genetics] has been provided not by a continuity of concepts and principles, though they did play a part, but by a continuing commitment to a set of procedures, goals, and problems" (1973, p. 1123).

The Variation and Selection of Mathematical Concepts

One general idea advanced by this chapter is the notion that variation and selection are at play in the development of mathematics. Additionally, according to Toulmin's definition of discipline, the understanding of how mathematics developed seems a reasonable fit. In *The Art of Algebra from Al-Khwarizmi to Viète: A Study in the Natural Selection of Ideas*, historian and mathematician Karen Parshall (1988) offers an account of how mathematical ideas *vary* according to social, historical, and idiosyncratic disciplinary factors and are *selected* based on a determination of their "fitness." She uses an evolutionary framework to explore the history of mathematics from a different, richer angle than is often the case. Whereas historians and mathematicians frequently think of mathematical ideas as developing in a linear progression with more recent and increasingly complex ideas logically building on prior ideas, Parshall argues that when viewed in this manner, the complex social dimensions of particular "mathematical climates" from any given time are compressed or underemphasized. The result is roughly akin to the impoverished version of mathematics as the neatly ordered discovery or recovery of preexistent Platonic objects or the development of logically deduced formal systems.

The introduction of the evolutionary frame seeks to contextualize and add complexity to the history of mathematics. In Darwinian fashion, Parshall starts by seeking to understand the environment: "Suppose that at any given time and place, we define the mathematical environment as the known body of mathematical facts, techniques, theories, and ideas together with the mathematicians who deal with them" (1988, p. 129). She goes on to explain that within a particular context

> every idea which presents itself, whether new or newly rediscovered, effects a change in the environment. Thus, the individual mathematician, by generating new ideas, by remaining ignorant of an idea, or by failing to absorb an idea, shapes the particular niche within which his or her own theories develop.
>
> *(p. 129)*

Parshall details how new ideas vary slightly from the ideas of a mathematician's recent predecessors and that these differences eventually lead to a variety of mathematical ideas from which the "more fit" will be selected. In this way, the particular intellectual atmosphere or environment from which a given mathematical idea was born becomes an important part of the evolutionary process:

> Viewed with respect to this kind of an evolutionary framework, what the modern mathematician and some historians might regard as the false starts, ill-conceived techniques, and imperfectly formed theories of the past, actually appear as intermediate steps in the evolutionary process of descent with modification.
>
> *(p. 130)*

Parshall's account offers a vivid picture of a populational understanding of concepts within the discipline of mathematics. At any given time there are many different ideas existent within a particular discipline. Each is tied to the history (both recent and distant) of the discipline as a whole as well as to the other ideas that are present in its local community. Like Toulmin's, Parshall's account seems more concerned with the relations of ideas within disciplinary boundaries and not so much of the relations between particular ideas and the problems of the wider community. Although the evolutionary account we are developing does seek to explore the relations between mathematics and our general quest for a better existence, there is no conflict here. The role of mathematics in human problem solving has also evolved over time. Whereas according to Kitcher, Dewey, and others, mathematics probably has its origins in helping individuals in their efforts to cultivate land, as mathematics developed it eventually reached a point where its potential for generalizability through abstraction from particular empirical situations was deemed beneficial.[31] Recall Kitcher's explanation of the role of the discipline of mathematics in our contemporary world. He explains that the way history has unfolded has left mathematicians with a "special role" to perform. They are

> licensed to devise new languages that relate in ways they find interesting and illuminating to the corpus they have inherited. The demarcation of that role itself represents a discovery about community inquiry, to wit that it is good for other investigations that the role be filled.
>
> *(2003, p. 411)*

Mathematics is still useful in helping humans solve their problems, it is just that rather than existing solely as a way to help individuals quantify their experience, it has evolved into a highly developed and specialized scientific discipline that helps to address a set of broadly conceived and themselves evolving ideas. While the preceding statement is quite close to Kitcher's account of the nature of mathematical knowledge, this chapter renders his account richer and more plausible.

An Evolutionary Philosophy of Mathematics

Dewey's insistence on the import of psychology to the development of mathematics provided a counter to the absolutist conceptions of mathematics as unsullied by the imperfect and psychology-ridden human mind. In grounding mathematical experiences in the context of our activities, Dewey's work also strikes a blow against relativistic and idiosyncratic accounts of mathematics forwarded by radical constructivists, as the results of mathematical inquiries can be judged by how well they meet the goals of the inquiry that was itself grounded in context; it is an *evolutionary* account. His detailed explanation of how humans come to develop concepts of number also served to enrich Kitcher's account of the empirical origins of mathematics. Likewise, Lakatos's approach can be used to enrich the historical dimensions of Kitcher's account through an exploration of the ways in which mathematics develops, both within the discipline and the classroom. The situated theorists contributed the notion that even in what is typically thought of as the highly formal and mental enterprise of mathematics, our thought is always tied to action. Furthermore, even in mathematics, our physical environment and social arrangements are key factors in its development. This contribution is critical, as it makes it difficult (if not impossible) to think of mathematics as one long flow of logical deductions (recall Kitcher's caricature of the mathematician, seated in his study, independently accessing the Truths that are mathematics). The inclusion of Toulmin and Parshall's work is intended to present a specific way in which the development of mathematical ideas can be thought of as evolutionary as well as to contend with the failure of Kitcher's account to suggest a mechanism for novelty.

While we do make specific philosophical arguments about the strengths and weaknesses of the various approaches presented in this chapter, we also hope that a general evolutionary gestalt begins to emerge from reading and thinking about the overall collection of ideas. In the event that our efforts to balance between precise argument and the building of a broader way of thinking have favored the latter, we have included a table at the end of this chapter designed to clarify what each theorist or theory specifically contributes to an evolutionary conception of mathematics.

A final note about our general aims and the applicability of evolutionary theory to philosophy of mathematics is in order. At this chapter's outset we explained that we are not trying to provide an account of the way that things "really are" regarding the nature of mathematics. While we have diligently endeavored not to overstate the connections between evolution and the development of concepts (mathematical, in this case), we do believe that there is worth in pushing the comparison a bit further. Toulmin asks and answers the following question: "How do historical entities maintain their coherence and continuity, despite all the real changes they undergo?" (1972, p. 356). Keeping in mind that the historical entities with which he is concerned are scientific disciplines—and our argument is that for the sake of Toulmin's analysis, mathematics can be considered a historical entity/scientific discipline—Toulmin's response eloquently strikes a balance between over- and underestimating the potential power of the evolutionary perspective outside of the realm of biology:

> Darwin's straightforward populational schema may not, in all cases, give the correct answer to this question, but that schema certainly provides one of the legitimate forms of answer. The balance between variation and selection within a population of constituent elements is, evidently, one of the possible processes by which historical entities preserve their transient identity. The task of exploring the implications of this populational schema in completely general terms—regardless of whether it will eventually be applied to organic species, or languages, or intellectual disciplines—is one that philosophers should be prepared to take seriously now. In this respect, the implications of an evolutionary approach to our theoretical problems— in the sense of populational, rather than progressivist approach—can take us beyond the limits of particular special sciences, and require us to reappraise our categories and patterns of analysis even on the most general philosophical level.
>
> *(Toulmin, 1972, p. 356)*

Thus, while not providing *the* account of how mathematics really is, the evolutionary perspective can foster *an* understanding of mathematics that will help us to deal with some of the problems endemic to our current situation. Toulmin's explanation of the power of his "populational" approach to get beyond the two choices of absolutism or relativism is a promising explanation of the worth of employing an evolutionary perspective to the philosophy of mathematics. Finally, the recognition of the ways in which this emerging evolutionary perspective is

TABLE 5.1 Summary of Contributions to an Evolutionary Perspective

Theorist/theory	Contributions
Phillip Kitcher	Provides general framework (from empirical origins through chain of knowers to today's discipline) Reintroduces empirical considerations (via Mill) Emphasizes historical nature of mathematical knowledge Develops the importance of the mathematical community Explains *how* mathematics develops
J.S. Mill	Recognizes the role of the physical world in shaping and constraining mathematics
John Dewey	Psychological explanation strengthens Mill's empirical account by emphasizing human activity Deepens Kitcher's account of how mathematics develops with practical account of *why* it develops (organism–environment interactions) Mediates the knowledge–belief distinction by viewing knowledge as coming from belief
Imre Lakatos	Provides detailed account of how Kitcher's mathematical community works Demonstrates how even the highly formal realm of mathematical community is inherently human-influenced Connected empirical methods of science and mathematics Demonstrates relevance of teaching and learning to mathematical development
C.S. Peirce	Bridges gap between Kitcher's origins and contemporary mathematics by showing the pragmatic/empirical dimensions of today's mathematics
Situated cognition	Combats Kitcher's rigid rationality with context/task-specific rationality Bridges gap between Kitcher's origins and present day mathematics by showing how mathematics is not always formal and disciplinary
Stephen Toulmin	Broadens understanding of rationality Specific employment of Darwinian mechanisms Populational approach accounts for novelty in mathematics
Karen Parshall	Enriches historical context of mathematics Employs variation and selection to explain development of mathematics Describes particular mathematical communities in populational terms

itself part of a larger historical entity (the rational enterprise that is philosophy of mathematics) and will endure to the extent that it is "selected" based on its ability to solve the problems that it addresses, gives rise to the hope that it offers a legitimate alternative to its competitors.

Our emergent evolutionary philosophy of mathematics, in addition to adding depth to the ways in which we can conceptualize mathematics and mathematical activity, also provides a way of thinking about mathematics that allows for inclusion in the broader social purposes of schooling. That mathematics is often thought about in ways that are radically different from the rest of school subject matter, both the absolutist orientation and the constructivist reaction to absolutism have made it difficult to connect school mathematics and non-math aims. Now that absolutism and constructivism have been described and critiqued and an evolutionary alternative has been developed, it is time to make explicit philosophical connections among mathematics, education, and democracy. To do so we next turn to Dewey's conception of an evolving democratic society.

Notes

1 Von Glasersfeld doubts the possibility of getting beyond individual idiosyncratic beliefs about mathematics. Ernest talks about objective knowledge, but his understanding of the term centers on public presentation and conventional agreement, not correspondence to some actual fact-of-the-matter.

2 Toulmin similarly considers the enterprise of predicting social futures. He cites (using other terms) the vastness of potential design space and speaks of "futuribles." "They are futures which do not simply happen of *themselves*, but can be *made to* happen, if we meanwhile adopt wise attitudes and policies" (Toulmin, 1991, p. 2). In our view, this is a clear and decidedly optimistic example of thinking about the role of contingency.

3 Russell and Whitehead's *Principia* is perhaps more suitable for Humphrey's thought experiment, as it uses math for math's sake rather than to describe the natural world. Of course, elsewhere we argue that there is no mathematics independent of human interaction, but certainly there are degrees of human involvement and the *Principia*'s formalism, once past the eminently human act of its creation, is relatively free-standing with regard to human influence.

4 Carl Sagan's novel *Contact* (1985) is a particularly well-developed version of this line of thinking, as virtually any reading of the ambiguous ending involves the recognition of mathematics as the way to come to understand our existence.

5 A shortcoming of Kitcher's is the gap he leaves between the pragmatic, empirical origins of mathematics and its highly abstract contemporary character. Later C. S. Peirce and the situated theorists will enter as a means to bridge this gap (Peirce by showing how formal mathematics is still somewhat concrete and empirical and the situated theorists by exploring the ways in which informal, concrete mathematics is just as rational as the formal discipline).

6 Although an exploration of Kitcher's notion of rationality is in order if his overall philosophy of mathematics is to be clear, we first need to explain his idea of the origins of mathematics in a bit more detail.

7 For example, *reductio ad absurdum* (proof by contradiction) was considered invalid by so-called "intuitionists" such as Luitzen Brouwer, briefly mentioned in Chapter 3. Despite the disagreement between Brouwer and Hilbert in the early twentieth century

about the validity of *reductio* arguments, these arguments are universally accepted by modern mathematicians and play a critical role in Wiles's proof. For further reading on the intellectual struggles over mathematical proof, see Part II, "The Origins of Modern Mathematics," of the *Princeton Companion to Mathematics* (Gowers, 2008).

8 By social practicalities we are referring to inquiries that come from outside of the community of professional mathematicians. These social practicalities created a need for the development of communities of mathematicians in the first place. Evolving social practicalities have also influenced the development of these professional mathematics communities.

9 Recall the criticism of Piagetian constructivism that used Gould's work decrying the false belief that evolution is inevitably headed somewhere specific and great.

10 We acknowledge that the "island" of IAS is an environment that is heavily manipulated, with its denizens having been preselected for particular traits and its chance encounters engineered by social customs, and in many important ways, it could be seen less "natural" than, say, the Galapagos Islands. That said, it is certainly still an environment, and ideas there will vary and be selected or not based on a variety of factors. Also, we are troubled by the elitism suggested by this example and we feel compelled to point out that when we say "groups of mathematicians," it is meant to be an inclusive, rather than exclusive, description of a local population. While such a group could be constituted by mathematicians and scientists at an Ivy League institution, they could just as legitimately be made up of community members working to solve a problem, or students in a classroom doing the same.

11 Although Dewey and McLellan co-authored the book, we will refer only to Dewey throughout this and subsequent chapters, as the philosophy of mathematics we attempt to recover is Dewey's.

12 This argument seems somewhat reductive, as it suggests that gross quantitative measurement can detail the quality of what is being measured (measuring a library solely by number of books, for example). Dewey presents a fuller explanation elsewhere in *The Psychology of Number* (pp. 42–44). Nevertheless, Dewey's perception of counting as measuring was critiqued in a review by Henry Burchard Fine, then a professor of mathematics at Princeton. Fine objected to the suggestion that the act of counting would first require a conscious choice of unit (e.g., "book" or "horse") and a subsequent iteration of that unit, repeatedly matching it to each element of a set. Dewey rebutted: "The whole point here is *under what circumstances* does one, not a mathematician or for mathematical purposes, count a group of horses" (Dewey, 1896, emphasis added). Later in this chapter, we explain Dewey's consideration of the importance of the context of any act of quantification.

13 This rationale could smack of the post-Darwinian belief that "ontogeny recapitulates phylogeny." While today, this is not a very attractive rationale, it was present in Dewey's time. Dewey's insistence that children are becoming agents in the disciplines as they learn about and participate in them seems a sufficient safeguard against the influence of recapitulation theory. Stephen Jay Gould discusses how school curricula in the late nineteenth century were reshaped to facilitate that belief (see Gould's *The Mismeasure of Man*, pp. 143–144). It does seem that the more one adopts a historical understanding of mathematics and advocates learning experiences that put children in the position to carry out mathematician-like inquiries, the more likely it is that the child's experiences will bear some relation to the history of the discipline. The upcoming section on Lakatos revisits this idea.

14 While pragmatism and Millian empiricism might seem, at first blush, to make quite strange bedfellows, it is worth noting that William James dedicated a seminal work, *Pragmatism*, to "the memory of John Stuart Mill from whom I first learned the pragmatic openness of mind and whom my fancy likes to picture as our leader were he alive today" (James, 1907/1978, p. 4).

15 The etymology of "cipher" is equally revealing: it is derived from the Arabic "sifr," meaning "zero" or "nothing."

16 Dennett writes of this in Chapters 5 and 6 of *Darwin's Dangerous Idea.*

17 The story is fictional in the sense that it is not an actual account of what transpired in any particular classroom, but it is certainly not mere whimsy, as Lakatos's story mirrors a similar dialogue between mathematicians that took place over several centuries. This dimension of the work will be developed in the following section.

18 A polyhedron is a three-dimensional shape comprised of polygonal faces. A regular polyhedron, such as a cube, has faces that are all congruent regular polygons and the same number of faces meet at each vertex. For more detailed descriptions and complete definitions, see Cromwell's *Polyhedra*, particularly pp. 205–210.

19 This formula means that for any convex polyhedron, subtracting the number of edges from the number of vertices and adding the number of faces will come to 2. A cube, for example, has 8 vertices (corners). Subtract its 12 edges (where two faces meet) to get −4. Finally adding the 6 faces does yield 2. It is likely that trial and error refers to the class looking at cubes, pyramids, and other polyhedra in hopes of finding the pattern that they eventually found.

20 Mathematical monsters, a term coined by Henri Poincaré in the late 19th century, are not merely incidental to proof and proving. Given their arresting character, they have the power to shape and reshape lines of mathematical argumentation. Elizabeth de Freitas describes the mathematical monster as "not quite paradoxical, but somehow unruly and uncooperative, and yet also a source of potential invention. Examples of monsters are everywhere in mathematics, as though the discipline itself were a breeding ground for them." She further describes Lakatos's view of mathematics as "a dialectical process of creating opportunities for monsters to be born and then redesigning the rules in order to banish them" (2016, p. 651). We concur, adding that rules may also be redesigned in order to adjust monsters.

21 Lakatos traces the roots of the formula back to Descartes (c. 1639). Interestingly, Descartes had written each of the pieces of the formula separately but Lakatos claims that he would never have seen any reason to combine them into what became Euler's formula.

22 Peirce is widely regarded as the first American pragmatist. Additionally, his influence on Dewey (and James) is well documented (Buchler, 1955, p. ix).

23 This idea is reminiscent of Arthur Bentley's (1954) "boundary of the skin" as traditional philosophy's "last line of defense."

24 While it is plain to see how Ernest's social constructivism (and its reliance on Wittgenstein and Lakatos's community of mathematicians) welcomes the situated focus on the social group, it is Dewey's work that meshes particularly well with Lave's interest in how informal mathematics possesses its own order and logic.

25 While Lakatos argued against formalism, it is true that his version of mathematics paid little attention to "everyday" problem-solving, instead focusing on the ways mathematicians solved mathematical problems.

26 We are following Lave's lead in this regard, both in her speech (recording, March 10, 1992) as well as the general tone of the introduction to Rogoff and Lave's book *Everyday Cognition.*

27 Not all situated theorists overemphasize socially constructed settings. See William J. Clancey's work for an example of situated theory that uses physical manipulations as a starting point. As a computer scientist working in the area of artificial intelligence, Clancey is acutely aware of the effects of the physical environment on simulating human behavior and also on actual human behavior and consciousness (1999).

28 Stability refers to the notion that, as many traditionalists/absolutists like to point out, 2+2=4 (and other mathematical "facts") will always be so. While the actual knowledge in language arts, social studies and even science are subject to revision, mathematics stands as

is. This seeming permanence and perfection is one reason why traditionalists are not willing to concede any ground to constructivists and other reformers in the "math wars."

29 Toulmin explains how Plato, Aquinas, Descartes, and Kant were all searching for a certain foundation of rationality, albeit in different directions (p. 44).

30 Toulmin uses Frege and Frege's contemporary R.G. Collingwood as paradigmatic cases of absolutism and relativism, respectively (pp. 57–74). While Frege's ideas have been touched upon here, Collingwood's have not. See Collingwood's *Essay on Metaphysics* (1972) and *An Autobiography* (1978) for a comparison of his ideas to that of Fregean/Russellian absolutism.

31 Peirce's thoughts presented earlier in this chapter are helpful in this regard. Peirce saw the mathematician as serving a specified and important role in the wider community, namely to help others (e.g., engineers, physicists, businesspeople) reframe their problems in more solvable forms (pp. 209–210).

References

Becker, H. (1972). A school is a lousy place to learn anything in. *American Behavioral Scientist*, 17(1), 85–105.

Bentley, A. (1954). The human skin: Philosophy's last line of defense. In S. Ratnor (Ed.), *Inquiry into inquiries: Essays in social theory* (pp. 195–211). Boston: The Beacon Press.

Boaler, J. (2000). Intricacies of knowledge, practice and theory. In J. Boaler (Ed.), *Multiple perspectives on mathematics teaching and learning* (pp. 1–17). Westport, CT: Ablex.

Brown, J., Collins, A., & Duguid, P. (1989). Situated cognition and the culture of learning. *Educational Researcher*, 18(1), 32–42.

Buchler, J. (1955). *Philosophical writings of Peirce*. New York: Dover.

Clancey, W. (1999). *Conceptual coordination: How the mind orders experience in time*. Mahwah, NJ: Lawrence Erlbaum Associates.

Collingwood, R. (1972). *Essays on metaphysics*. Chicago: Henry Regnery Company.

Collingwood, R. (1978). *An autobiography*. New York: Oxford University Press.

Cromwell, P. (1997). *Polyhedra*. New York: Cambridge University Press.

de Freitas, E. (2016). Number sense and the calculating child: Measure, multiplicity and mathematical monsters. *Discourse: Studies in the Cultural Politics of Education*, 37(5), 650–661.

Dennett, D. (1995). *Darwin's dangerous idea: Evolution and the meanings of life*. New York: Simon & Schuster.

Devettere, R. (1973). Review of *Human understanding*. *International Philosophical Quarterly*, 13, 449–452.

Dewey, J. (1896). Psychology of number [Letter to the editor]. *Science*, 3(60), 286–289.

Dewey, J. (1910). The postulate of immediate empiricism. In *The influence of Darwin on philosophy and other essays* (pp. 1–19). New York: Henry Holt.

Dewey, J. (1910). The influence of Darwin on philosophy. In *The influence of Darwin on philosophy and other essays* (pp. 226–241). New York: Henry Holt.

Dewey, J. (1916). *Democracy and education*. New York: The Free Press.

Dewey, J. (1920/1967). *Reconstruction in philosophy*. Boston: Beacon Press.

Dewey, J. (1929). *The quest for certainty*. New York: Minton, Balch, & Company.

Ernest, P. (1991). *The philosophy of mathematics education*. Bristol, PA: The Falmer Press.

Fosnot, C. (1996). *Constructivism: Theory, perspectives and practice*. New York: Teachers College Press.

Frege, G. (1884/1960). *The foundations of arithmetic: A logico-mathematical enquiry into the concept of number.* J. L. Austin, Trans. (2nd ed.). Evanston, IL: Northwestern University Press.

Gould, S. (1996). *The mismeasure of man.* New York: W.W. Norton.

Gowers, T. (Ed.) (2008). *The Princeton companion to mathematics.* Princeton, NJ: Princeton University Press.

Hein, P. (1966). On problems. *Grooks.* Cambridge, MA: MIT Press.

Hilbert, D. (1964). On the infinite. In P. Benacerraf & H. Putnam (Eds.), *Philosophy of mathematics: Selected readings* (pp. 134–151). Englewood Cliffs, NJ: Prentice-Hall.

Hilbert, D. (1967). The foundations of mathematics. In J. van Heijenoort (Ed.), *From Frege to Gödel: A source book in mathematical logic, 1879–1931* (pp. 464–479). Cambridge, MA: Harvard University Press.

Hull, D. (1973). A populational approach to scientific change. *Science*, 182, 1121–1124.

Humphrey, N. (1987). Scientific Shakespeare. *Guardian* (London), August 26.

James, W. (1907/1978). *Pragmatism: A new name for some old ways of thinking.* Cambridge, MA: Harvard University Press.

Kant, I. (1963). Prolegomena to any future metaphysics which is to be a science. In *Kant* (G. Rabel, Trans.). London: Oxford University Press.

Kessler, G. (1980). Frege, Mill, and the foundations of arithmetic. *The Journal of Philosophy*, LXXVII(2), 65–79.

Kitcher, P. (1977). On the uses of rigorous proof. *Science*, 196, 782–783.

Kitcher, P. (1980). Arithmetic for the Millian. *Philosophical Studies*, 37, 215–236.

Kitcher, P. (1983). *The nature of mathematical knowledge.* New York: Oxford University Press.

Kitcher, P. (1988). Mathematical naturalism. In W. Aspray & P. Kitcher (Eds.), *Minnesota studies in the philosophy of science, volume XI: History and philosophy of modern mathematics* (pp. 293–325). Minneapolis: University of Minnesota Press.

Kitcher, P. (2003). Giving Darwin his due. In J. Hodge & G. Radick (Eds.), *The Cambridge companion to Darwin* (pp. 455–476). Cambridge: Cambridge University Press.

Kuhn, T.S. (1962). *The structure of scientific revolutions.* Chicago: The University of Chicago Press.

Lakatos, I. (1976). *Proofs and refutations.* Cambridge: Cambridge University Press.

Lave, J., Murtaugh, M., & de la Rocha, O. (1984). The dialectic of arithmetic in grocery shopping. In B. Rogoff & J. Lave (Eds.), *Everyday cognition: Its development in social context* (pp. 67–94). Cambridge, MA: Harvard University Press.

Lerman, S. (2000). The social turn in mathematics education research. In J. Boaler (Ed.), *Multiple perspectives on mathematics teaching and learning* (pp. 19–44). Westport, CT: Ablex.

McLellan, J., & Dewey, J. (1895). *The psychology of number and its applications to methods of teaching arithmetic.* New York: D. Appleton & Company.

Mill, J. (1843/1967). *A system of logic, ratiocinative and inductive, being a connected view of the principles of evidence and the methods of scientific investigation* (8th ed.). London: Longman's, Green and Co.

Ortiz, D. (1999). Categorical community. *Stanford Law Review*, 51, 769–806.

Parshall, K. (1988). The art of algebra from Al-Khwarizmi to Viète: A study in the natural selection of ideas. *History of Science*, 26, 129–164.

Peirce, C. (1898). Logic of mathematics in relation to education. *Educational Review*, v(XV), 209–216.

Ratner, S. (1992). John Dewey, E.H. Moore, and the philosophy of mathematics education in the twentieth century. *Journal of Mathematical Behavior*, 11, 105–116.

Rockmore, D. (2005). *Stalking the Riemann hypothesis*. New York: Knopf Doubleday.

Rogoff, B., & Lave, J. (1984). *Everyday cognition: Its development in social context*. Cambridge, MA: Harvard University Press.

Sagan, C. (1985). *Contact: A novel*. New York: Simon and Schuster.

Secada, J. (2001). Historiography. In L. Becker & C. Becker (Eds.), *The encyclopedia of ethics: Vol. 2* (pp. 683–685). New York: Garland/Routledge.

Singh, S. (1997). *Fermat's last theorem*. New York: Fourth Estate.

Thackray, A. (1974). Review of *Human understanding*. *British Journal for the History of Science*, 7, 80–81.

Toulmin, S. (1972). *Human understanding: The collective use and evolution of concepts*. Princeton, NJ: Princeton University Press.

Toulmin, S. (1991). *Cosmopolis: The hidden agenda of modernity*. New York: The Free Press.

6

DEMOCRATIC MATHEMATICS EDUCATION

In a preschool classroom, a group of children expresses the need for another work table, a duplicate of a table already in their classroom. They ask for help from a local carpenter, who tells the children that he needs the measurements. One child suggests they measure the table with their fingers; others use their hand spans. After many attempts, they reason that choosing a longer unit would be more efficient. They try everyday objects such as books and ladles and discover that as long as the object is smaller than the table, it will work as a tool to measure the table. In fact, objects prove far easier to manipulate than parts of the body. Eventually, they select Tommaso's shoe as the official unit of measurement. They determine that the table is six and a half shoes long and three shoes wide, and they are impatient to tell the carpenter. But someone insists there is still a problem: the carpenter will need Tommaso's shoe. This strikes Tommaso as impractical, and the children realize they must translate the length of the shoe in a way the carpenter will understand. After many days of tinkering with meter sticks and measuring tapes, they finalize their specifications in centimeters and communicate them to the carpenter in a letter and accompanying drawing (Castagnetti & Vecchi, 1997).

★ ★ ★

High school students are challenged to determine the ideal placements of bus stops along a particular city route. After reviewing the specific constraints of the problem, the students must decide what other variables are pertinent to the context (e.g., typical walking velocity, time needed for boarding and disembarking) and then devise a plan for data collection. Eventually, they realize that greater distances between stops mean longer walking distances for passengers, while shorter distances between stops mean a longer travel time overall—a different kind of inconvenience altogether. The tension between these and other factors helps students see that determining an optimal distance between stops requires making difficult choices. They must decide which factors matter the most, and to whom, and at what cost (Kaiser & Stender, 2013).

Work tables. Bus stops. The objects of inquiry in these classroom vignettes are strikingly ordinary. They are inconspicuous until, quite suddenly, they are not. The protracted investigation into measuring a table arose from a need articulated by preschool students, whose tenacity is due in large part to their pressing need for the problem to be resolved. Admittedly, we have not adequately represented the role of the teacher-researchers, who immediately recognized the potential (mathematical and otherwise) of the claim, "We need another table," and who artfully questioned and scaffolded the children throughout the life of the project. Still, we feature this investigation because of what it can tell us, and indeed what the children learned, about mathematics as a remarkably useful tool with the power to help realize change. Here, school mathematics is not an act of compliance but a collaborative action marked by intention, impediments, invention, and consequences.

For the high school students, the bus stop problem did not originate with students' own questions or observations. It was crafted by researchers interested in the relationship among mathematical modeling, autonomous learning environments, and the role of the teacher. Nevertheless, the problem is engaging, relevant to students' lives, and builds to a kind of crescendo. It takes as its focus something students might see or experience every day and not think worthy of much thought (let alone mathematizing), but it turns out to have a gripping complexity, both mathematically and practically. Small adjustments have surprisingly large impacts. Quantities become especially meaningful, reflecting facets of human experience that oblige students to reevaluate their model.

These vignettes can prime our thinking about what we mean by democratic mathematics education and help ground this work as a philosophy of practice. Having projected images of real classroom scenarios, we take the opportunity now to restate that democratic mathematics education is where the nature of mathematics, its teaching and learning, and the broader purposes of schooling can and should meet.

In this chapter, we sketch a pragmatic conception of democracy that starts with Dewey, includes some critique, and then contends with the critique by introducing the thoughts of some modern democratic theorists who draw inspiration from pragmatism.[1] We conclude with a summary of what a democratic mathematics education offers and requires. One important thing to bear in mind throughout is that Dewey saw all learning as necessarily tied to social context, and this idea has found purchase in most content areas. Mathematics education, for reasons articulated in earlier chapters of this book, has been particularly averse to the inclusion of social contexts to its learning. For democratic mathematics education to work, the social has to be recognized as part and parcel of mathematics learning.

Recall that McLellan and Dewey (1895) placed the origin story of mathematics within the human need to measure in an effort to live better and more efficient lives. That Dewey linked even mathematics—a discipline so often seen as entirely separate from human interests—to human values and aims compels us to develop a Deweyan philosophy of mathematics education that considers broader social and political contexts. In other words, forging a Deweyan philosophy of mathematics education requires that we first understand what Dewey meant by democracy.

Dewey's Conception of Democracy

To Dewey, social and political contexts were not distinct. In fact, he was less focused on technical matters of how democratic government ought to be structured, turning instead to the relationships that citizens ought to have with each other, and how these relationships connect individuals, the groups they form and re-form, and broader social aims and values.

Dewey saw it as useful to think of democratic societies as collections of smaller groups or "publics" and that the health of a democracy can, in part, be judged by whether these smaller publics are getting better at working within and between groups: "in any social group … we find some interests held in common, and we find a certain amount of interaction and cooperative intercourse with other groups" (1916, p. 83). Briefly stated, internal cohesion and external interaction are two elements necessary in democratic societies. Internal cohesion is present when societal direction emerges from multiple and varied points of common interest. External interaction is exemplified as groups previously isolated from one another (based on class, race, education, ideology, nationality, etc.) are able to interrelate and reconstitute their social behaviors and norms based on these relationships.

Within this frame of macro- and micropublics, Dewey positions a crucial feature of democracy: it balances individual and societal flourishing. That is, a healthy democracy values both the development of individual capacity and the subsequent demand that citizens give back to society. Dewey saw the development of individual interests with consideration to others as having great promise to break down the social barriers existing in his time. This idea is encapsulated in what is probably his most famous declaration about democracy:

> A democracy is more than a form of government; it is primarily a mode of associated living, of conjoint communicated experience. The extension in space of the number of individuals who participate in an interest so that each has to refer his own action to that of others, and to consider the action of others to give point and direction to his own, is equivalent to the breaking down of those barriers of class, race, and national territory which kept men [sic] from perceiving the full import of their activity.
>
> *(1916, p. 87)*

"Individual" and "others" do not constitute a dualism but instead function as perpetually interacting bodies, so to speak, that necessarily guide and shape one another. It is through this continuous association that actions are evaluated and, importantly, learning occurs. Democracy *is* the ongoing communication, not the result of it. It is a wide way of living, a commitment to being in constant relation to one another so that each and all may flourish.

Schooling and the Aims of Democracy

Dewey posited that democratic communities are inextricably devoted to systematic, public education:

> A society marked off into classes need be specially attentive only to the education of its ruling elements. A society which is mobile, which is full of channels for the distribution of a change occurring anywhere, must see to it that its members are educated to personal initiative and adaptability. Otherwise, they will be overwhelmed by the changes in which they are caught and whose significance or connections they do not perceive. The result will be a confusion in which a few will appropriate to themselves the results of the blind and externally directed activities of others.
>
> *(1916, p. 88)*

Dewey saw himself as responding to traditional forms of education, ones "marked off into classes," that have their roots in Ancient Greece. Dewey acknowledged Plato's positive contributions to educational philosophy, particularly Plato's emphasis on the links between the cultivation of one's natural abilities and the health of a society. Dewey agreed that the primary aim of education is to help individuals discover their skills, talents, and interests—to develop them and use them in effective ways—but he was fiercely critical of Plato's prescriptive educational model because in it, skills and talents served as limiting factors, inhibiting interrelation and reconstitution of social habits. Plato called for a group of philosophers who would sort people into groups according to ability. People preoccupied with the most basic human pursuits would provide manual labor. Educated individuals lacking strong capacity for reason would be guardians of the peace and citizen-subjects. Finally, those possessing an education of the highest order (abstract, universal knowledge) would lead and govern (Plato, 1928). Dewey was disturbed by the similarities he saw in industrial society and, to some degree, in much of the educational philosophy and psychology of the early to mid-20th century.

Dewey (1916) criticized narrow views of intelligence that focused on efficiency and productivity to the exclusion of social factors. Capitalists controlled the aims of industry and provided workers with training focused on efficient skill development, determined by an external initiative and most certainly devoid of social interrelation. Dewey considered activity detached from aims, under the purposes of another, as anti-democratic. The capacities of workers were being developed and providing a social return, but workers' lack of self-determination and purpose perpetuated class divisions and undermined democratic principles.

Anti-democratic education that disconnected ends from means started early and persisted throughout history, up to and well past Dewey's time. Dewey indicated that "The notion that the 'essentials' of elementary education are the three R's mechanically treated, is based upon ignorance of the essentials needed for realization of democratic ideals" (1916, p. 192). The mechanical treatment of education presupposes that the end of education is social-capital return and earned income. Assuming that education's end is to "make a living" or provide social return detached from significant action robs the individual of a meaningful present existence because the end to which they work is unrecognizable to them.[2]

Dewey, of course, articulated different aims for a robustly democratic education which did not limit intelligence to a mechanical participation in means based on ends that exist outside of or beyond the activity. Instead, the aim of education must be present within existing situations. Intelligent activity allows the learner to modify actions based on identified ends. This is important for democratic societies because only in understanding the original aim can the individual reflect upon existing conditions and change the target. Rigid, external aims prevent teachers and students from understanding that ends are experimental, ongoing, and inextricably related to particular contexts. Furthermore, they are future means, not ultimate ends. By freeing activity through the legitimate use of aims, the potential for changing conditions exists (1916). Personal initiative—even care and passion—and adaptability are key components of an education that measures up to democratic principles. This philosophy of education provides a foundation for the readjustment of social habits, the second element that points to democracy.

Dewey saw the connections between ends and means as necessary for human activity to be free and undertaken out of desire and not imposition: "Every means is a temporary end until we have attained it. Every end becomes a means of carrying activity further as soon as it is achieved" (1916, p. 106). The underpinning of the forms of education Dewey was working against is that there are ultimate, abstract ends which can be externally imposed. Plato saw these abstract and ultimate ends as the highest forms of education and thus exclusive to leaders and governors. This rational form of epistemology, which purports a hierarchy of knowledge and class, separates knowledge from activity and citizens from one another.

To Dewey, in a vibrant democracy, such rigid ends are undesirable and, eventually, untenable. The very notion of predetermined ends leaves no room for *indeterminate* situations, let alone individuals who might make judgments and take some sort of action. In other words, predetermined ends obviate problems and intelligent agents. A robust democratic education cannot focus on predetermined ultimate ends, it must cultivate the ability to propose aims, construct means to achieve those ends, and to evaluate when ends need to be adjusted in light of changing circumstances. Although Dewey's way of thinking about the ends of education has had some impact on many facets of the curriculum,[3] mathematics education has remained largely resistant to such ways of thinking. We argue that this is mostly related to the widespread belief that mathematical knowledge is different in kind from other forms of knowledge. Given this predominant way of thinking, there is no way for Dewey's link between democracy and education to extend to mathematics education. The previous chapter's development of mathematics in evolutionary terms, coupled with this articulation of a Deweyan mathematics education, make possible the meaningful connection of democratic and mathematics education.

Democracy and Mathematics Education: Dewey's Laboratory School

While Dewey did not provide an explicit and well-developed discussion of how his naturalized philosophy of mathematics relates to his broader democratic project, reconstructing a Deweyan argument is both possible and, we argue, urgently useful. A starting point for this reconstruction is *Democracy and Education*'s chapter on "Educational Values." It provides sustained discussion of the unfortunate tendency for subject matter to be broken up and justified according to specific sub-values. In a rare, direct consideration of mathematics education in the book, Dewey states:

> As matter of fact, such schemes of values of studies are largely but unconscious justifications of the curriculum with which one is familiar … Mathematics is said to have, for example, disciplinary value in habituating the pupil to accuracy of statement and closeness of reasoning; it has utilitarian value in giving command of the arts of calculation involved in trade and the arts; culture value in its enlargement of the imagination in dealing with the most general relations of things; even religious value in its concept of the infinite and allied ideas. But clearly mathematics does not accomplish such results, because it is endowed with miraculous potencies called values; it has these values if and when it accomplishes these results, and not otherwise. The statements may help a teacher to a larger vision of the possible results to be effected by instruction in mathematical topics. But unfortunately, the tendency is to treat the statement as indicating powers inherently residing in the subject, whether they operate or not, and thus to give it a rigid justification. If they do not operate, the blame is put not on the subject as taught, but on the indifference and recalcitrancy of pupils.

> *(1916, p. 254)*

Here Dewey critiques the idea that the subject itself has inherent value independent of our social activities. It is possible that such a sentiment is easily accepted in many domains, but the absolutist outlook is very influential and it sets up mathematics as valuable specifically because it transcends human, social contexts.[4] Dewey is attempting to inextricably link curriculum to human contexts—a radical move when applied to mathematics. His choice to place mathematics in the values chapter of *Democracy and Education* serves to move the reader beyond the narrow skill, fact, and understanding-based aims typical with mathematics education and into the realm of broader purposes. If social contexts and broader aims are considered, then democratization of mathematics is possible.

Looking to sources beyond *Democracy and Education* can provide clarity as to what Dewey advocated regarding classroom practice. The University of Chicago Laboratory School's annual reports (1898–1899) document how mathematics was taught there and how its teaching was conceptualized. School activities were intentionally selected and developed to support essential themes of science and history. They were also highly social in nature and, importantly, connected to student interest. For example, one teacher describes the circumstances leading to students' work with ratios and division of fractions:

> [The students] needed to know the ratio of revolution of the large to the small wheel in spinning. They got the diameter of the large wheel and worked out the circumference of it as approximately 3 and 1–7 times the diameter,[5] which they had used in finding the contents of a globe last fall. The numerical work involved a division of fractions, and as they were rusty in this, an hour was spent in practice.
>
> *(University of Chicago, Box 1, Folder 25, p. 39)*[6]

Having spent time investigating the mechanisms of a spinning wheel, students applied familiar concepts and yet untried procedures to a new problem, one that was born of inherent, organic interest.

Students' mathematical work also provided some access to the historical development of mathematics and focused on understanding as opposed to raw, rote skill development. Catherine Camp Mayhew and Anna Camp Edwards's account of Dewey's Laboratory School, *The Dewey School*, can help us to see how Deweyan math instruction is related to DME. Mayhew and Edwards were both teachers in the laboratory school affiliated with Dewey and the University of Chicago and in a sense, one can read *The Dewey School* as a set of demonstrations of the ideas laid out in *Democracy and Education*, and as an important way to see Dewey's ideas in action.[7] Mayhew and Edwards start with children's early mathematical experiences (from concepts of "much/many" to "how much?/how many?") and move through progressively more complex ideas, arising from student interests and experience. For example:

In early practice in weighing, the idea of an equation was given as a way of representing the drawing power of the earth on different objects. The earth-pull on an object on one side of the balance is equal to the earth-pull on a number of objects on the other (i.e., the weights) and was represented symbolically in series as x (the unknown weight) = a + b + c.

(Mayhew & Edwards, 1936, p. 343)

Mayhew and Edwards make explicit their belief about the role of the teacher in scaffolding mathematical development: "At all times, [the teacher] must watch for the psychological time and place to introduce symbols and the opportunity to formulate generalizations" (p. 345). Just as the spinning wheel provided an authentic context for fraction computation, here equations are connected to the action of a two-pan balance, thus readily relating symbols to their referents. As a result the students likely developed a robust and relational, not superficial, understanding of equivalence.

We should clarify that a Deweyan democratic approach does not justify teaching a deep understanding of equivalence by asserting that equivalence itself has some special, intrinsic value. Rather, what matters is the way in which equations (and by extension, inequalities) successfully describe the world and our experience in it. They are useful to us, just as standard units of measure were useful to the preschool children, mentioned at the onset of this chapter, in communicating their needs to the carpenter. Even amid the ubiquitous and admirable appeal for learning mathematics with understanding, we suggest that mathematics is not to be learned simply for its own sake. Again, we turn to Mayhew and Edwards's description of the culture and overarching tenets of mathematics as realized at Dewey's Laboratory School:

Always, at whatever stage, number was taught not as number but as a means through which some activity, undertaken on its own account, was rendered more orderly and effective. In this way it afforded insight into the ways in which man actually employs numerical relations in social life.

(p. 345)

Laboratory School teachers also applied Dewey's historical contextualization of mathematics, seamlessly integrating this history into student interests:

A child becomes interested not only in the origin of the symbols for number, but in measurement units of all kinds. Through his [sic] appreciation of primitive man's use of sun and moon as time measurers, he takes interest in his own calendar. This generally happens first in connection with something that intimately concerns him, such as his own birthday or the first day of school. His ability to read the clock, which has been progressing slowly, is usually perfected about this time. His ideas continually enlarge through his daily experience. Whereas once he measured his garden by the number of his own paces, he now begins to use the yardstick or the ruler to find how his patch compares with the length of his neighbor's.[8]

(p. 342)

Laboratory School mathematics helped students solve problems not primarily rooted in the content of math class. Integrating mathematical activities into other subject areas (e.g., math in the service of science) encouraged recognition in children that they can act on the world and, perhaps more importantly, it could foster a desire to do so.

Integrating academic disciplines and attending to students' interests were hallmarks of Dewey's Laboratory School and have endured as tenets of progressive educational reform. They may, in fact, sound familiar to our modern ears. It is important to remember that Deweyan educational reform is inseparable from his broader political theory and comprehensive social vision. His laboratory for teaching and learning was not just an expression of his pedagogic creed, it was a community in which his democratic ethos was enacted daily:

> When the school introduces and trains each child of society into membership within such a little community, saturating him [sic] with the spirit of service, and providing him with the instruments of effective self-direction, we shall have the deepest and best guaranty of a larger society which is worthy, lovely, and harmonious.
>
> *(Dewey, 1899, p. 44)*

The democratic school, therefore, is a place where an emphasis on both individuation and commitment to community, or "spirit of service," infuses the educational and, hence, social landscape.

Critiques of Democratic Theory/Pragmatism Applied to DME

Throughout this project we have injected critique as a means of identifying and questioning the assumptions, some more conspicuous than others, of a range of theories and philosophical orientations. Although they are, or perhaps because they are, our chosen approaches, democratic theory and Deweyan pragmatism are equally in need of critique. Grounded in a pragmatic/evolutionary framework and explicitly related to the broader social aims of schooling, our vision of school mathematics can help, we believe, to redress persistent inequities both in and out of the math classroom. Our conviction that school mathematics can enact and cultivate Deweyan democratic ideals is sustained by a sense of unsentimental and clear-eyed hope[9] and by trust in the promise of democracy, or rather what democracy makes possible. Certainly, we find Dewey's ideas useful as a way to think about math and math education, but we also recognize that it is up to us to interrogate and reconstruct what we have found inspirational in Dewey as our contexts inevitably change. His thinking is of somewhat limited use to us now because at best he is simply not of our time, and as Dewey himself would assert: ideas belong in context. At worst, Dewey was too nearly agnostic on issues of equity, diversity, and particularly, race. Some contemporary pragmatists who have focused on race have refracted Dewey's thinking through a different prism, one that shines light on new possibilities in Dewey's thought and in pragmatism. Cornel West's critique of Dewey's thinking as overly optimistic can serve as a point of departure. Eddie Glaude Jr.'s subsequent response showing how evolutionary perspectives can link contingency and human agency is crucial to our contemporary reconstruction of Dewey. To Glaude, if there were no contingency (and all the uncertainty and tragedy that accompany it) in our human lives, then there would be no need for human agency. West's multifaceted critique of Deweyan pragmatism and particularly Glaude's reconstruction has implications for how we can conceive of community and to the obligations that we have to one another.

West and the Tragic

West is a towering figure in contemporary philosophy and an influential public intellectual. His contemporary version of pragmatism begins with a Deweyan foundation and is modified in ways that make it particularly useful in considering some of today's problems, especially the relationship between race and democracy. In *Democracy Matters* (2004), West draws on historical sources to remind us of our democratic potential. He also links democracy and efforts toward social justice by engaging with prophetic religion. West points out problems and contradictions in our democracy, yet he also has faith in the possibilities of democracy and the moral commitments and visions it entails, which "fortify the soul, empower, and inspire" (p. 15). In a passage that starts with an invocation of democracy's role in social change, West explains that "democracy is always a movement of an energized public to make elites responsible—it is at its core and most basic foundation the taking back of one's powers in the face and the misuse of elite power" (p. 68). He goes on to articulate a Dewey-influenced overview:

> democracy is more a verb than a noun—it is more a dynamic striving and collective movement than a static order or stationary status quo. Democracy is not just a system of governance, as we tend to think of it, but a cultural way of being.
>
> *(p. 68)*

In linking action and agency, West reminds us of pragmatism's particular promise as a way to animate efforts designed to make our democracy live up to its ideals. West adds to this general action-agency orientation a focus on the Black experience and history of struggle. West positions the resulting prophetic prag-matism as helping pragmatism to be able to contend with what he sees as its primary challenge, namely that it lacks a sense of the tragic.

In earlier work, particularly *The American Evasion of Philosophy*, West leveled another critique at pragmatism, one which he traces back to its origins in the thought of Ralph Waldo Emerson. West saw Emerson as providing three major influences on pragmatism, two of which he saw as primarily positive: 1. rejecting the push for epistemological certainty and 2. what Robert Westbrook summarizes as Emerson's "romantic vision of human agency and possibility" (2005, p. 204). The third was to elevate the American preoccupation with individualism and to allow it to merge with a particular strain of elitism. According to West, Emerso-nian self-reliance led to willful disconnection from larger social groups and a point of view foregrounding "individual conscience along with political impo-tence, moral transgression devoid of fundamental social transformation, power without empowering the lower classes ... and human personality disjoined from communal action" (1989, p. 40).

In West's characterization of Emerson's three major influences on pragmatism, we can see how the first two support the vision of democratic mathematics education that is beginning to emerge. The rejection of epistemological certainty helps us move beyond a traditional absolutist outlook about mathematics and mathematics education. Likewise, the foregrounding of agency helps to solidify the importance of human beings in the creation and development of mathema-tical knowledge. The third influence, individualism that leads to elitism, is parti-cularly troubling in the context of mathematics education, as overly individual accounts of both the creation of mathematical knowledge and the learning of mathematics abound. Individualism is a feature of both the absolutist and con-structivist traditions—it is rife throughout absolutism and present in con-structivism, primarily in the psychological constructivist realm. Of course, elitism is also a long-running problem in mathematics education, as evidenced in the way mathematics is often held up as a pure form of knowledge and in how decontextualized mathematical acumen (usually in the form of standardized tests) is used as a proxy for general intelligence and ability.

While West sees Dewey as attempting to contextualize Emerson's politics in a way that is more inclusive, community-minded, and robustly democratic, he also sees a need to reconstruct Deweyan thought for our modern social problems, parti-cularly the problem of racism. West's introduction of the Black Christian prophetic tradition's foregrounding of history and family/community is a way that he attempts to counter the individualist strain of pragmatist thought. According to this tradition, one is always inextricably linked to those who came before and their struggle to live and to provide possibility for those who will come after. Thus, recognizing the power of history—tragedy and all—and the communities that are forged partially in response to tragedy to shape the vision of an egalitarian future for our plural diverse society is at the core of West's pragmatism.

We next turn to Eddie Glaude, Jr., in order to further develop a community-oriented pragmatism as a way to enhance thinking about democracy and the possibility of a democratic mathematics education. Another benefit of Glaude's entrance into the conversation is that, like us, he centers the evolutionary strains of Dewey's thought in order to reconstruct a version of pragmatism that is up to the task of contending with some problems in our 21st-century context.

Glaude, Jr.: From Contingency to Responsibility and Community

Eddie Glaude, Jr., a former student of West, is another contemporary pragmatist who sees the worth of using Dewey's broader philosophy and his thinking about democracy as a tool to consider today's vexing social problems. While he and West have largely complementary visions of how a revised pragmatism might contribute to democratic renewal—particularly with regard to race, racism, and democracy—Glaude, parting ways with West, sees Dewey's pragmatism as containing some ability to understand and contend with tragedy and conflict. Interestingly, given the central place we have given to the evolution–Dewey connection and our application of this connection to our philosophy of mathematics education, Glaude sees Dewey's focus on evolution as the means by which pragmatism engages with the tragic dimensions of life.[10]

Glaude claims that locating a sense of the tragic in Dewey's thought requires some reconstructive work. He takes this task on by focusing on the central role of change in Dewey's notion of experience and the difficulties of our moral choices in this world of perpetual flux: "This understanding of the precariousness of the world of action is absolutely crucial for understanding the tragic vision implicit in Dewey's pragmatic philosophy of action" (Glaude, 2007, p. 22). Humans respond to the perennial uncertainty of the world by engaging in practical action and making choices. And though we may seek certainty and equilibrium in the face of a dilemma, there is never any guarantee these will be realized. Tragedy, Glaude argues, is "a moral phenomenon" (p. 20) because despite our best efforts to improve our lives or the lives of others, despite our actions and choices, the outcome may or may not be beneficial to us or to those for whom we care. The contingency of outcomes is the contingency of human life. But rather than render us hapless victims of a cruel and capricious world, uncertainty is instead the lifeblood of human agency. The way to proceed is to turn our attention to our lived experiences and actions, not pursue a Truth greater than ourselves or some predetermined end. Only then can we make a claim not just to our own agency in shaping and improving the world, but to our responsibility as actors in it:

> This connection to the future forms the primary basis for responsibility. For in the efforts to secure our world for our children and ourselves, we employ methods that generate foresight. We make moral and political prognoses

with an eye towards securing and expanding for future generations we cherish.

(Glaude, 2007, p. 24)

Responsibility has both temporal and social dimensions: it reaches forward to the future and outward to the community. The word "responsibility" more precisely describes our commitment to associated living. It is an enactment of the sentiment, "I care about what happens to you and yours." We have seen that democracy is more about the process of its own becoming than about some institutional end, and Glaude's call for responsibility is a timely update to this Deweyan conception of democracy. It is not just interaction among individuals and groups that matters but our obligation to one another. Though the world gives up none of its stochastic indifference, exerting its influence "for weal or woe" as Glaude states, human agency is fortified by compassion and community.

Stitzlein's "Habits of Democracy"

Responsibility could fairly be considered a habit, in the Deweyan sense of the word. In *Habits of Democracy* (2014), Sarah Stitzlein clarifies what Dewey means by "habit," distinguishing it from the customary sense of habit as an unconscious impulse formed over time through repetition. She describes a more willful inclination, one that informs how we act and reason: "Habits … shape and precede the generation of ideas. They provide us with know-how, 'working capacities' that help us know how to act in the world" (2014, p. 63). Most importantly, habits are mutable. That is to say, they develop and change as warranted by new situations and by social and environmental interactions. So while habits drive the reasoning we deploy during the process of inquiry, inquiry itself has a bearing on those habits. The mutability of habits is essential to developing democratic sensibilities, just as it is essential to individual human development. Indeed, *un*democratic education necessarily precludes the readjustment of social habits by limiting or even prohibiting social interrelation.

In keeping with Dewey's conception of democracy as a mode of associated living rather than a remote destination, schools, as sites of assembly and shared experience, are in a position to help cultivate what Stitzlein calls "habits of democracy." Habits of democracy are not rigidly defined citizenship goals but sensitivities and ways of being/acting that may themselves be adapted to meet unknown future challenges. Mathematics, as it is in other ways, is a subject often overlooked in discourses of citizenship and character education. This too is a missed opportunity. Stitzlein's five identified habits of democracy are infused throughout the discussions about DME in the following chapters and we invite the reader to imagine how mathematical inquiry in particular might help nurture these proclivities:

Citizenship as shared fate stresses the importance of caring for others and for one's community. This habit is connected to our sense of social responsibility, which directs our actions and choices so that we take into consideration the fate and well-being of all group members.

Collaboration and compromise are dispositions related to working and thinking together and include the actions of co-construction and mediation. Collaboration and compromise can be marked by discord but also by a commitment to working toward the common good.

Deliberation involves intentional discussion and listening and requires a genuine willingness—even an eagerness—to hear alternative ideas and points of view. An openness to ideas is not synonymous with "neutrality." It stems instead from an authentic curiosity and a resistance to ideological stagnation.

Analysis and critique are akin to critical thinking and sharpen our ability to question, to dissent, and to identify injustice in our world. Analysis and critique may lead to action, with an eye towards improving institutions and circumstances for oneself and for others. This habit has elements of both healthy skepticism and meliorism.

Hope is the understanding that democracy itself is changeable. Different from optimism, hope is the habit that actually impels us to take action. In other words, a habit of hope means we fundamentally believe that our actions can have some effect.

In the short final section we take the discussion of Dewey's political philosophy and how it relates to schooling and DME and consider it in light of the contemporary critique and extensions from West, Glaude, and Stitzlein, and we distill it into a brief description of DME.

Tenets of Democratic Mathematics Education

DME engages an evolutionary/pragmatic approach to thinking about the nature of mathematical knowledge. Within this framework, mathematics is evaluated on the basis of how well it meets the goals of inquiry, which is itself context-bound. Instead of belonging to an external "perfect" realm or existing as constructions of particular individuals or groups, mathematics is connected to the actual lives humans lead and is shaped by its uses, or functions. DME embraces a broad understanding of rationality such that mathematics is not restricted to formal and abstract domains. It is a set of tools we use to contend with the changing circumstances of our lives. DME steadfastly rejects a social-capital view of the mathematical "worth" of individuals and must itself be democratic in its enactment.

The idea that mathematics is, to some degree, "shaped by its uses" might seem strange or a controversial claim, but thinking about it through the lens of Deweyan democratic theory can help to render it understandable. Certainly, much of the content of mathematics class is remarkably durable but, as noted earlier, there is much to gain in thinking about mathematical knowledge as different in degree but not in kind from other forms of knowledge. When children "do" math in school, the effects of treating math as permanent, unchanging, and finished can make genuine democratic education impossible. Inquiry is a hallmark of DME, and inquiry requires live investigations of which the outcomes are not entirely predetermined. The non-foundational, evolution-influenced philosophy of mathematics that we have developed makes it possible for the door to genuine inquiry to be opened in math class, thus making Deweyan reconstruction possible. This opening is subtle, as of course most genuine inquiries in school mathematics confirm the work of previous mathematicians. Recall Lakatos's description of a class learning about the relationships among the vertices, edges, and faces of certain three-dimensional figures. The students engage with the problem earnestly. As mathematicians, not even fledging mathematicians, they work through the problem and in doing so Lakatos shows how their inquiry mirrors the history of mathematicians' related work. They learn how to be a part of a community of inquirers and all that entails. Perhaps most importantly, in presenting the corresponding history of the arguments related to the problem, Lakatos develops a version of mathematics that is unfinished. It is a messy process replete with contradictions and revisions. Lakatos links the students' inquiry to this unfinished work. The key point here is that when students are engaged in genuine inquiry they are not accepting a preformed set of truths and, consequently, they are learning to inquire and not just accept or consume. Learning to inquire carries with it much of what is needed in a democracy.

DME is strongly participatory in nature, emphasizing both personal and group initiative. Grounded in social contexts, problematic situations relate to the human story in some way. DME rejects the notion of predetermined ends and instead cultivates the ability to propose aims, construct means to achieve those ends, and evaluate when ends need to be readjusted in light of changing circumstances. It proceeds through social relations, where actions, choices, models, and solutions are repeatedly modified based on social engagement and on mutually identified and negotiated ends. Thus, it generates an iterative process of adapting solutions and models in response to interactions with others and with the physical world. DME also highlights the role of moral choice in mathematical decision making. How we use mathematics inevitably involves decisions about what to include and how to represent facets of human experience, and these decisions have consequences. We bear responsibility to others when we use math. In fact, DME positions the ethical use of mathematics as one important way to build solidarity with others.

In sum, DME engages students in relevant, contextualized social inquiries designed to develop students' abilities to pose problems, to test solutions, to evaluate the work based in part on how well the work solves the initial problems. This work is intended to be done in concert with others and to foster agency, care, and responsibility. In the following chapter, we consider some of the ways in which these philosophies play out in the classroom. After presenting math classroom activities that are absolutist and constructivist in orientation we provide a description of an activity that possesses the ethos and many of the characteristics of DME.

Notes

1 See Stemhagen and Smith (2008) for an earlier effort to link Dewey's political thought to his mathematics education.
2 Dewey's critique of an economic rationale for education would be echoed nearly 100 years later by mathematics educator Paola Valero, among others. On the topic of mathematics education in particular she cautions, "I think that we run a serious risk of reducing the meaning of mathematics education to education for the qualification of a submissive workforce. In such context, we end up educating not a human being but a *homo œconomicus*. This person does not need to be a thinking person, or a rational being, but an economic exchange being" (Valero, 2018, p. 114).
3 We don't want to overstate Dewey's influence on the day-to-day operation of public education. We believe, à la Labaree (2010) and Lagemann (1989), that the efficiency progressives and behaviorists have had much more lasting (and often pernicious) influence on schooling. That said, Dewey is often cited in academic work in most curricular areas, but not so often in mathematics education.
4 In spite of the National Council of Teachers of Mathematics' constructivist reforms, the "math as certain pure knowledge" idea runs deep. From the appeal of math class as the one place where you know when you get the right answer to the use of math achievement as a proxy for general intelligence, absolutism stubbornly persists.
5 We take this to mean "3 and 1/7 times the diameter."
6 For an in-depth look at the practices of Dewey's Laboratory School, see Laurel Tanner's comprehensive 1997 publication, *Dewey's Laboratory School: Lessons for Today*.
7 Of course, the activities of the Laboratory School predate *Democracy and Education*. The development of Dewey's ideas and the related practices can be thought of as a real-life manifestation of Dewey's conceptualization of theory and practice as two phases of activity. The ideas for Dewey's earliest education writings were put into action in the Laboratory School and *Democracy and Education* represents his best thinking in light of the tentative results of his theory in action.
8 Placing the development of mathematical ideas in historical context is only one part of naturalizing/humanizing the subject. The Laboratory School attached math skills and knowledge to its history *and* to the live interests of students.
9 Kathy Hytten (2019) and others call this democratic hope. Drawing on Dewey, West, and others, she characterizes it by active connections between visions of a better world and concrete actions and habits that can help bring it about. This form of hope is also necessarily social. Hytten explains that it requires open-mindedness and "learning from others and truly listening and attending to diverse perspectives" (p. 11).
10 For more on the importance of Darwin's influence on Dewey given the task of remaking pragmatism for today's problems, see Melvin Rogers's *The Undiscovered Dewey: Religion, Morality, and the Ethos of Democracy* (2009).

References

Castagnetti, M., & Vecchi, V. (1997). *Scarpa e metro: I bambini e la misura. Primi approcci alla scoperta, alla funzione, all'uso della misura* [Shoe and meter: Children and measurement. First approaches to the discovery, function, and use of measurement]. Reggio Emilia, Italy: Reggio Children.

Dewey, J. (1899). *The school and society*. Chicago: The University of Chicago Press.

Dewey, J. (1916). *Democracy and education*. New York: The Free Press.

Glaude, E.S., Jr. (2007). *In a shade of blue: Pragmatism and the politics of Black America*. Chicago: The University of Chicago Press.

Hytten, K. (2019). Cultivating democratic hope in dark times: Strategies for action. *Education and Culture*, 35(1), 3–27.

Kaiser, G., & Stender, P. (2013). Complex modelling problems in co-operative, self-directed learning environments. In G.A. Stillman, G. Kaiser, W. Blum, & J.P. Brown (Eds.), *Teaching mathematical modelling: Connecting to research and practice* (pp. 277–293). New York: Springer.

Labaree, D. (2010). How Dewey lost: The victory of David Snedden. In D. Trohler, T. Schlag, & R. Osterwalder (Eds.), *Pragmatism and modernities* (pp. 163–188). Rotterdam, The Netherlands: Sense.

Lagemann, E.C. (1989). The plural worlds of educational research. *History of Education Quarterly*, 29(2), 185–214.

Lakatos, I. (1976). *Proofs and refutations*. Cambridge: Cambridge University Press.

Mayhew, K.C., & Edwards, A.C. (1936). *The Dewey school: The laboratory school of the University of Chicago 1896–1903*. New York: D. Appleton-Century Company.

McLellan, J., & Dewey, J. (1895). *The psychology of number and its applications to methods of teaching arithmetic*. New York: D. Appleton & Company.

Plato (1928). *The republic* (B. Jowett, Trans.). New York: Charles Scribner's Sons.

Rogers, M. (2009) *The undiscovered Dewey: Religion, morality, and the ethos of democracy*. New York: Columbia University Press.

Stemhagen, K. & Smith, J. (2008). Dewey, democracy, and mathematics education: Reconceptualizing the last bastion of curricular certainty. *Education and Culture*, 24(3), 25–40.

Stitzlein, S.M. (2014). Habits of democracy: A Deweyan approach to citizenship education in America today. *Education and Culture*, 30(2), 61–86.

Tanner, L. (1997). *Dewey's laboratory school: Lessons for today*. New York: Teachers College Press.

University of Chicago. Laboratory Schools. Work Reports, 1898–1934. Special Collections Research Center, University of Chicago Research Library. https://www.lib.uchicago.edu/e/scrc/findingaids/view.php?eadid=ICU.SPCL.LABSCHOOLREPORT

Valero, P. (2018). Human capitals: School mathematics and the making of the homo œconomicus. *Journal of Urban Mathematics Education*, 11(1&2), 103–117.

West, C. (1989). *The American evasion of philosophy: A genealogy of pragmatism*. Madison: University of Wisconsin Press.

West, C. (2004). *Democracy matters: Winning the fight against imperialism*. New York: Penguin Press.

Westbrook, R. (2005). *Democratic hope: Pragmatism and the politics of truth*. Syracuse. NY: Cornell University Press.

PART IV

Enactment: Philosophy of Mathematics in and out of the Classroom

7

PHILOSOPHIES OF MATHEMATICS
Consequences and Classroom Expressions

Implications

We now turn to some of the possible educational consequences of adopting particular philosophies of mathematics. The earlier discussion of the math wars is fraught with cases of individuals and groups who have taken philosophical positions largely out of a distaste for opposing ways of thinking. It is not the adoption of a constructivist philosophy of mathematics education that necessarily causes problems; it is instead when one uses the constructivist argument as a means to dogmatically react against mathematical absolutism that trouble ensues. Rather than base classroom instruction on the generally beneficial tenet that children learn best when they have a stake in constructing (making) their own knowledge, the reactive constructivist stubbornly argues that children always and only make their own knowledge and that there is no mathematical knowledge that is "found" (recalling Rorty's found/made distinction). This dogmatic outlook can force its adherents to defend the extreme position that practicing discrete skills has virtually no place in mathematics class.

Examples of the opposite problem—adopting an overly absolutist understanding of mathematics—can, we fear, be found simply by thinking back to many of our experiences as students in mathematics class, or even by looking in some of today's classrooms. An overreliance on "drill and kill" methods where students learn a particular skill and practice it until proficient is one example. Any class in which mathematics is treated as a body of Truth or skills that is transferred from the text or teacher to our heads and that we are to do our best to master is probably an example of the decision makers (teacher, curriculum specialists, textbook makers, etc.) working, at least implicitly, with some form of an absolutist understanding of mathematics.

It is our suspicion, however, that the justification for most rote learning is not the manifestation of a carefully reasoned philosophical position so much as it is the result of other influences. Beyond the immediate and obvious influences, the ways in which teachers teach mathematics are influenced by any number of factors, such as habit, political orientation, a belief in the importance of the mental rigor of a skills-approach to mathematics, to name just a few. An underlying philosophy is certainly not the only or perhaps even the primary shaper of classroom practices. This does not weaken the value of this project, however, as we argue that philosophical considerations *should* play a larger role in teacher decisions. Basing pedagogical decisions on how well they are in tune with a well-considered philosophical orientation helps to bring means and ends into harmony. Teachers possessing clear ideas about the nature of the subject and its teaching and learning are more likely to be able to teach with direction and conviction—making purposeful choices about how to teach and how to engage with students. When teachers are not encouraged to think about the nature of their subject matter and the broader purposes of why they are teaching it, they become more susceptible to basing pedagogical decisions on the possibly counterproductive reasons stated above. Though philosophies do not deductively determine behavior, it does seem that the way one conceives of the nature of the subject matter probably does play a hand in how one thinks the subject ought to be taught. The democratic approach developed in this project is an alternative to the polar extremes of absolutism and constructivism as philosophies of mathematics, with the hope that such a philosophical reconsideration can have a positive effect on the teaching and learning of mathematics.

In what immediately follows, we consider what can emerge in classrooms where absolutist and constructivist philosophies are followed to the exclusion of all others. We present an activity that uses DME as its philosophical orientation. The theoretical construct of mathematics as possessing a front and back enriches the discussion of the promise of DME in mathematics classrooms. Finally, we offer a short discussion of how DME can help avoid the problems inherent in classrooms subscribing to overly absolutist or radically constructivist understandings of mathematics.

Limitations of the Exclusivity of Finding or Making

Absolutism

It needs to be stated that on some levels, mathematical absolutism is an effective philosophy of mathematics education. Just as there are different forms of mathematical absolutism (recall the earlier descriptions of Platonism and formalism), there are different ways in which absolutism might play out in the classroom. The existence of a formal, relatively unchanging curriculum that is to be conveyed tends toward mathematical absolutism insofar as mathematics is portrayed or understood to be a body of knowledge or set of skills that are separate from human activity. While we believe that a static curriculum does carry with it a set of potentially problematic consequences, the stability of such a formalized conception of mathematics can help ensure that individuals will be able to communicate with each other mathematically and that a certain level of rigor is the standard in our schools. Furthermore, a static but rigorous and carefully thought out curriculum can help students learn, over a period of years, much of what the history of mathematics has taken centuries to produce.

In addition to a standard curriculum, Platonist strains of absolutism encourage the development of the notion that mathematics is an activity in which individuals try to come to understand and learn about objects and entities that exist in a non-spatiotemporal realm. Earlier, we discussed that the seemingly constructivist contemporary pedagogical technique of employing manipulatives in the mathematics classroom can actually be a way in which teachers attempt to teach children about the qualities of mathematical objects only existent in an abstract realm. According to this conception of mathematics education, teachers see their role as helping children come to know a form of truth that is permanent, unchanging, and unsullied by our human lives.[1]

A formalist version of absolutism might lead to a classroom where the acquisition of particular sets of skills is the central aim of class. Furthermore, mathematics might be depicted as actually being the sets of symbols and their "correct" manipulation. For example, an algebra class taught in this manner would be based primarily on working to ensure that the students learn how to carry out procedures related to solving equations for given variables, factoring, distributing terms, etc.

As an example, consider quadratic functions, a long-standing algebra topic. Quadratic functions are polynomial functions of the form $y = ax^2 + bx + c$, where $a \neq 0$. The graph of a quadratic function is a parabola in which its characteristics, such as location and shape, depend on the values of a, b, and c. A math classroom tending towards absolutism may introduce quadratic functions in a decontextualized manner, perhaps beginning with the very definition we have supplied, and then unfold procedure by procedure (e.g., graphing quadratic functions, recognizing their different written forms, factoring and solving quadratic equations, and so on). The occurrence of quadratic relationships in a variety of real-world problems, from throwing a ball to maximizing rectangular areas, matters only to the extent that students solve them efficiently and accurately. In such an environment, contextual problems would typically be stripped of any worldly complexity, free from the kinks and knots of everyday life. By the time students encounter these problems they are effectively *already solved*, and it is the student's task to execute the correct procedure and "find" the answer.

At its worst, absolutism leads to the idea that mathematics only involves discovering and coming to understand that which exists independently of human activity. This strong "mathematics is found" position attempts to rule out the importance of a student's psychological knowledge construction (at least as anything that is relevant in any way other than as a ladder which is to be thrown away once the rungs of understanding have been climbed). Additionally, a consequence of an absolutist conception of mathematics might be to de-emphasize or even ignore the reasons why mathematical techniques were first created/discovered as well as any of the ways in which mathematics might be useful in our contemporary lives.

Absolutist mathematics classrooms generally tend to promote a passive version of education. Mathematics, rather than being thought of as a tool to aid in the achievement of some human end, is taught as something to be worshipped as if it were an artifact in a museum. To complete the metaphor, we are cordoned off from mathematics, allowed only to look at and revere it, but never to recognize how it has grown out of and is inextricably linked to our actual lives. The development of the ability to use mathematics to contend with "real world" situations is not of central import in most absolutist classrooms, as to most absolutists, mathematics did not originate in this real world.

We expect that many facets of the absolutist mathematics classroom should be quite familiar, as we hypothesize that most teachers operate under a Platonist outlook, be it overt or latent.[2] Whether Platonist, formalist, or some other form of absolutism, there is a body of literature that supports our suspicions. Alan Schoenfeld (1992) notes that an outcome of seeing mathematical knowledge as the set of mathematical facts and procedures one can reliably and correctly use

> is that instruction has traditionally focused on the content aspect of knowledge ... The route to learning consists of delineating the desired subject-matter content as clearly as possible, carving it into bite-sized pieces, and providing explicit instruction and practice on each of those pieces so that the students master them.
>
> *(p. 334)*

Describing general characteristics of the absolutist classroom runs the risk of lapsing into caricature, particularly given our philosophical criticism of the tendency to view mathematics in this manner. Likewise, asking the reader to remember negative experiences as a mathematics student and claiming that these experiences were probably the result of an absolutist outlook is not only too simplistic, it is probably spurious. There is no essential way to be absolutist and there is no thinker who is completely absolutist.[3] Still, it is fair to say that an absolutist mathematics classroom is principally concerned with understanding a preexistent structure of mathematics and demonstrating mastery of related skills. The overall picture of a classroom influenced by this type of philosophy of mathematics education is one where the problems and answers are predetermined, as are—to a large extent—the ways to solve the problems.

Constructivism/Fallibilism

Mathematical fallibilists have worked to present a more complex and human-oriented version of the nature of mathematics. Emboldened by the publication of the National Council of Teachers of Mathematics' *Principles and Standards for School Mathematics* and by the ensuing development of standards-based curricula such as *Investigations in Number, Data and Space* and *Mathematics in Context*, constructivist mathematics education experienced something of a golden age beginning in the early 1990s. At their best, non-absolutist versions of mathematics serve as a much-needed corrective to absolutist outlooks. The adoption of a fallibilist philosophy of mathematics can "link mathematics with people, with society, with history" (Hersh, 1997, p. 238). That said, it is also true that the adoption of a rigidly fallibilist conception of mathematics (in defiant opposition to mathematical absolutism) can produce its own set of problems. Furthermore, one way to envision the place of fallibilism/constructivism in the wider philosophy of mathematics education conversation is that it is a way of thinking that emerged to help contend with some of the problems that existed as a result of a rigidly absolutist outlook. Thought of in this way it can be seen as growing out of the absolutist outlook, although not necessarily outgrowing some of the deeply entrenched and problematic dualistic ways of thinking. Whereas absolutism is primarily concerned with transmitting knowledge of static ideas and training students to perform set algorithms, constructivism is characterized by a focus on the development of concepts. That is, absolutism is primarily procedural and constructivism is mostly conceptual.

In the original edition of a book still widely used today to prepare elementary and middle school mathematics teachers, John Van de Walle sets the constructivist tone in the preface:

> This book is about the challenging and rewarding task of helping children develop ideas and relationships about mathematics. The methods and activities that you will find throughout the book are designed to get children mentally involved in the construction of those ideas and relationships. Children (and adults) do not learn mathematics by remembering rules or mastering mechanical skills. They use the ideas they have to invent new ones or modify the old. The challenge is to create clear inner logic, not master mindless rules.
>
> *(1990, p. vii)*[4]

Constructivist mathematics teachers tend to focus on the mental processes by which students come to develop mathematical understandings. Furthermore, for many constructivists, these mental processes are conceived of from a Piagetian perspective. So, whereas the structure of the subject matter is of primary concern to many absolutists, constructivists tend to concern themselves with mental structures.[5] This helps us to understand why constructivist teachers tend to ask more open questions than their absolutist counterparts, as they need to know about the thought processes of their students. This preoccupation with what is happening inside of the child's mind is captured with excerpts from Van de Walle's explanation of the principal characteristics of teaching developmentally:

> It is realizing that children, not teachers or books, give meaning to ideas and procedures … it is encouraging children to talk about concepts and relationships … it is using manipulative models as a major tool to create linkages between conceptual and procedural knowledge.
>
> *(p. 18)*

Constructivism in Practice: Confrey's Example

In "What Constructivism Implies for Teaching," mathematics educator Jere Confrey gives an account of an "outstanding teacher with constructivist beliefs" (1990, p. 113). In the previous section we mentioned that presenting a fictional example of absolutist or constructivist methods runs the risk of creating a straw man or lapsing into caricature. Consequently, it is important to note that Confrey's article details the methods of a constructivist, as determined by Confrey (a self-proclaimed constructivist), and published in a constructivist publication.[6] We detail some of the methods employed by this constructivist teacher and explain the rationales for their implementation.

Confrey claims that helping students develop their reflective processes is a critical component of mathematics education, going so far as to say that "reflection is the bootstrap for the construction of mathematical ideas" (p. 116). She identified three steps or levels in this process and gave examples of the ways in which the teacher taught accordingly. Confrey explains:

> for students to modify and adapt their constructions, they must: (1) encounter a situation that they experience as personally problematic; as a roadblock to where they wish to be; (2) act to resolve the problematic, often using multiple forms of representation and (3) assess the success of their action in resolving the problematic or determine what problematic remains.
>
> *(p. 116)*

The teacher dealt with the first level by getting the students to rephrase the question, asking questions such as "What is the problem?" or "What does the problem say?" The second level was confronted through the students' explanation of what they were doing. Confrey reports on one such interaction involving the teacher and a student as she confronts a problem involving a comparison of two fractions with unlike denominators (13/5 and 21/10). The student is using the technique of drawing pictures to compare the fractions and they are comparing this attempt with an attempt to graphically depict a comparison of 2/3 and 5/7:

TEACHER: How did we do the fifths and the tenths?
STUDENT: But those were in proportion.

TEACHER: What do you mean, "in proportion?"

STUDENT: Not proportion; they were equal. At least, um, I mean they weren't equal, but—I know what I mean.

TEACHER: I know what you mean too, but now you'll have to tell me.

STUDENT: I mean I can't think of a word. I mean five is half of ten; therefore we divided the fives in half. It would be just like adding five more. I don't know how to explain it. I mean you have like five parts and you divide it in half, and it was like double. But like if you took three things and divided it in half, you'd have six things and not seven. Therefore, I mean you'd have a different problem.

(p. 117)

The third level primarily involved a defense of the work on the part of the student. Confrey explains that the teacher looked for explanations that "adequately fit the student's interpretation of the problem and the methods and strategies she had constructed" (p. 117). While there is discussion of other facets of this teacher's pedagogy, developing the powers of reflection is at the core of this teacher's understanding of what teaching mathematics is all about.

Confrey is sensitive to the critique of constructivism that depicts it as a way of thinking which has no means to determine a correct answer and is antithetical to mathematics education, as in mathematics there is generally a "right answer." The fact that the justification phase (level three) finds the teacher mostly worrying about how the student's evaluation of her work fits within her own thought processes and not some outside standard of truth may understandably have prompted Confrey's concern. In a section titled "Adherence to the Intent of the Materials," Confrey defends the constructivist's ability to teach according to a particular agenda and to have a means to determine whether learning is taking place. Of the opponents to constructivism, Confrey remarks,

> They mistakenly believe that a constructivist teacher lacks a specific agenda for what is to be learned in the classroom. Such a characterization did not apply to this teacher. He was committed to a particular view of mathematics learning and found many opportunities to share this with the students.
>
> *(p. 122)*

The particular view that the teacher possessed was a determination to "have the students come to see that one can make sense of fractions using pictures, and that the algorithms for rational numbers can be seen as actions on those pictures" (p. 122). The bulk of his teaching involved determining where students were in terms of their development through listening to verbalizations of their thought processes and helping them relate their work to his pictoral model of representations of fractions.

In contrast to primarily absolutist mathematics classrooms, the focus of instruction in this constructivist classroom was on helping students develop conceptual understandings of mathematics. Whereas a typical absolutist might teach students a step-by-step procedure for comparing fractions, most likely by first finding a common denominator, this constructivist teacher chose to help students understand what it means to compare fractions through the use of pictures. While this teacher did not do so, some constructivist teachers will teach the algorithms, after the fact, as shortcuts that can be used once conceptual knowledge is present. Either way, this particular example of constructivist teaching, while thoroughly student-oriented, does seem to suffer from an inability to have a clear way to determine whether or not the mathematics that has been learned is working.

Comparing the constructivist and absolutist classrooms shows that in contrast to the absolutist's tendency to focus on demonstrating proficiency in skills and understanding of discrete pieces of information in an effort to get the single correct answer, constructivists tend to have a primary concern with mental structures and processes and the demonstration or expression of conceptual understandings of mathematical relationships. With absolutism, the problems, answers, and paths to solution are all precast. With constructivism only the problems are usually predetermined and the worth of the answers and means of arriving at the answers are largely dependent upon the current stage of development of the student's mental structures.

Possibilities of the Democratic Approach

One of the authors served as an instructor in a state-run seminar for Algebra I teachers with the specific objective of helping teachers find and integrate websites designed to highlight the uses of algebra. On one of the featured websites, students could enter an equation, and the computer not only solved for desired variables but also displayed each step on the way to the solution. Inevitably there were two reactions from the participating teachers. One only slightly exaggerated response was that the technology would be the downfall of humankind as we know it because students no longer had to "do any math." The other reaction was that it was a wonderful and powerful teaching tool. Both groups, though, failed to recognize the opportunity the website offered.

The first group feared the technology threatened the entire enterprise of conveying the body of knowledge that is mathematics. The second group recognized its potential to help students better understand the meanings of the specific skills that make up the Algebra I curriculum. But the tool, variations of which abound on the Internet today, instead invites us to reconsider just what the Algebra I curriculum should be. In other words, the current Algebra I curriculum is based on the idea that the set of skills required to solve equations *is* mathematics. The advances in technology demonstrated by this website offer the opportunity for us to stop spending the bulk of our time in mathematics class learning to calculate and more time actually engaging in mathematical activity—in using mathematics as a means to deal with situations taking place not in a textbook, but in our lives.

To illustrate this point, participating teachers were given the following fictional scenario: "A former Algebra I student has finished your class and is enjoying her summer. She is outside one day and recognizes a situation that could benefit from algebraic modeling. She determines the equation, runs inside, and enters the equation into her computer. The website spits out the solution for the variable she specified. Next, she takes the solution and applies it to her initial situation and it makes her life a little bit easier." The teachers were then asked if they would feel they had done their job as a teacher if this fictional girl were a student of theirs. Unanimously, the teachers admitted that they would feel good about such a situation. But when it was pointed out that the girl did not engage in much activity that is considered fundamental to the Algebra I course in our state (according to state standards) there was recognition that maybe mathematics is more than just the petrified curriculum and that today's Algebra I course (as is much of the K–12 mathematics curriculum) *is mostly about learning to do what the computer did and not what the girl did.*

The advent of the computer is just one in a series of unending changes to our circumstances that can cause alterations to our notion of what mathematics is and, consequently, what mathematics education ought to be. DME requires that we be willing to broaden our conception of mathematics to include the human activity from which it originated as well as the human interactions that allow it to continue to develop. Furthermore, if this perspective is to stay true to its evolutionary view of mathematical knowledge, then we must also acknowledge that this is not the final word on how to think about mathematics, as changing circumstances both within the discipline and in our broader social arrangements will most likely require reassessment of our understandings of mathematics. Prior to a presentation of how mathematics education can benefit from this perspective, it might be helpful to consider in more detail what is meant by our call for a broader understanding on the part of mathematics educators as to what is to count as mathematics. In this regard, Reuben Hersh provides a useful metaphor, referring to the front part and back part of mathematics.

Hersh's Metaphor: Letting Students in the Kitchen

In *What Is Mathematics, Really?*, mathematician Reuben Hersh offers an interesting and different way to think about mathematics. Hersh explains that math can be thought of as divided into two areas, front and back. The idea, an application of sociologist Erving Goffman's work (1973), is that the finished product of mathematicians belongs in the well-ordered and more-or-less highly polished front of mathematics while the back is the area where mathematicians are busy engaging in their messy but often fruitful practices. He uses the analogy of a restaurant. The front of a restaurant is the dining room and the back is the kitchen. In the dining room everything is to appear orderly and under control. Those in the front are not privy to all that goes on behind the scenes in order to create the seamless experience of dining in the front. Hersh explains math in these terms:

> The front and back of mathematics aren't physical locations like the dining
> room and kitchen. They're its public and private aspects. The front is open
> to outsiders; the back is restricted to insiders. The front is mathematics in
> finished form—lectures, textbooks, journals. The back is mathematics among
> working mathematicians, told in offices or at café tables … Front mathe-
> matics is formal, precise, ordered, and abstract. It's broken into definitions,
> theorems, and remarks. Every question either is answered or is labeled:
> "open question." At the beginning of each chapter a goal is stated. At the
> end of each chapter, it's attained. Mathematics in back is fragmentary, infor-
> mal, intuitive, tentative. We try this or that. We say "maybe," or "it looks
> like."
>
> *(1997, p. 36)*

It is a common belief that the front part of mathematics is all that exists. Our call
to expand what counts as mathematics, continuing with the Hersh/Goffman
metaphor, is an invitation for all students to leave the well-ordered dining room
and to see what's cooking in the kitchen (as well as *how* things are cooking, to do
some cooking themselves, and to learn to evaluate the quality of what has been
cooked). In the preface to *What Is Mathematics, Really?* Hersh explicitly states his
beliefs about mathematics and the goal of his work: "I show *that from the viewpoint
of philosophy* mathematics must be understood as a human activity, a social phenomenon, part of human culture, historically evolved, and intelligible only in a
social context. I call this viewpoint 'humanist'" (1997, p. xi).

Hersh stresses that we do mathematics first and philosophize about it later. He
does not deny the seeming banality of declaring that mathematics is a human
activity that takes place in the context of a society, but failing to recognize the
importance of mathematics' sociohistorical context is, he feels, the source of the
intractability of many of the problems of the philosophy of mathematics.

Mathematizing

At the root of both Hersh's front/back discussion and our own call for a broader
conceptualization of mathematics is an attempt to include human activity in context
in discussions regarding the nature of mathematics. Simply tacking application problems onto procedural instruction, as is frequently the case, will not accomplish this
goal. Likewise, in mathematics education, explaining how mastering the mathematics curriculum will help out in the "real world" will also not suffice (although it
is, we contend, a step in the right direction). Instead, this project's evolutionary
perspective requires that human agency must be folded into any sensible understanding of what mathematics is and from where it came. While it is obvious how
the contingent, human facets of mathematics differ from the tenets of absolutism, it is
also a departure from constructivism, as constructivists still mostly espouse understandings of mathematics whereby the primary context is the mathematics class. In

other words, the only contingency that constructivists tend to acknowledge is that which is created in individual minds or in social groups and not that which is due to the changing contexts within which users of mathematics actually find themselves.

In her thoroughly situated account of mathematics in practice, *An Ethnographic Study of the Mathematical Beliefs of a Group of Carpenters*, Wendy Millroy explains that in the English language we really have no word that adequately describes the activity of creating or engaging in mathematics:

> The only way in which we can refer to this multifaceted human phenomenon is via the noun "mathematics," which seems to refer to a static body of knowledge that already exists. We need a word to express the concept of an active process of creating and using mathematical ideas or tools. I shall use the verb "mathematizing," which embodies action and refers to the experience of creating and using mathematical ideas.
>
> *(1992, pp. 9–10)*

Thus, without mathematizing there would be no mathematics. In "Consensus and Coherence in Mathematics: How Can They Be Explained in a Culturalistic View?," Susanne Prediger (2002) presents a similar argument. Although her argument is more concerned with philosophy of mathematics and the discipline of professional mathematics than it is with mathematics education, her ideas have strong educational implications. Here she responds to the idea that mathematics is not an appropriate domain for sociological analysis because, according to Prediger, her opponents claim that mathematics is "immune against human influence" (p. 3). According to Prediger:

> We must emphatically reject this view of mathematics, because in this account, crucial areas of mathematical activities are ignored. The entire process of mathematization (i.e. the question how initial non-mathematical problems are to be translated into mathematics), the concept formation, and the development of theories as well as the criteria of relevance of research questions are missing. How are mathematical concepts found? What influences the process of concept formation? How does the community decide whether a problem is adequately mathematized? Which factors affect the development of a theory? Who decides about the relevance of questions or theorems? In all these fields, the contingent character of mathematics is much more evident than in a simple limitation on proving.
>
> *(p. 3)*

Millroy and Prediger's accounts of the human element in mathematics—Millroy's situated mathematizing by non-mathematicians and Prediger's account of human involvement in the formalization of such mathematizing—combine very nicely to offer a description of how mathematics educators operating with a democratic perspective might conceptualize their subject matter.

DME in Practice: Weighted Formulae

Taking Millroy and Prediger's conceptualizations to the next level brings us to a mathematics activity that embodies this understanding. On the surface, the purpose of this activity is to help students understand and use weighted formulae.[7] It was also designed to spark discussion regarding the objectivity of numbers and even the limits of the enterprise of quantification. In this way, this lesson also offers a meta-level point of entry for class discussions regarding the nature of mathematics. It begins with a discussion of whether we can objectively rank or judge performances in different domains. The classroom participation grade given by many teachers or essay-test scoring are good points of departure, as students are not just familiar with such practices but are invested in whether or not they are fair. In a guided discussion about class participation, middle school students can establish some quantitative measures of participation (e.g., frequency of hands raised, frequency of being called upon, frequency of giving the "right" answer, not distracting other students as measured inversely by the number of times a student has been reprimanded for talking, etc.). Students fairly quickly see that even once quantified, human influence has not been purged from the system, both in terms of the selection of criteria as well as the ways in which those criteria will be measured. For example, the maker of the system (teacher) ultimately gets to decide whether to include frequency of volunteering versus frequency of actually getting called upon. Middle school students will be only too happy to point out that teachers are not always fair in calling on students. Additionally, human influence is evident in the manner in which the criteria, once chosen, will be rated. In other words, if the teacher decides to reward "correct" answers, how will this be judged? Is partial credit given? What about effort?

There is a bolder project on the horizon that will eventually find each student creating a rating system of their own. They will use a weighted formula to evaluate any phenomenon they find worthy of such attention, say cars, clothes, or even the weather. Teachers can model the process by developing a weighted formula for another phenomenon of interest to middle school students: the rating of popular songs. They begin by selecting particular criteria and ignoring others in their evaluative formula. At every stage, they model not just mathematics but their role as decision maker. For example, the particular criteria (and variables) might be: lyrics (L), catchiness (C), musicianship (M), and coolness (K). In constructing a formula, the coefficients (weights) could total 10, and when considering a particular song, a 0–10 scale could be used for assigning values to each criterion. This combination yields a score range of 0–100. Altering either the sum of the coefficients or the rating scale will change the final range and demonstrate the elasticity and sensitivity of the mathematics. Also in the spirit of mathematical exploration, students can independently weigh the variables, creating a classroom full of slightly different formulas that may be compared. One example of how a formula might look is: $R = 4L + 3C + 1.5M + 1.5K$.[8] Next, a particular song is chosen, and students score each criterion, say on a scale from 0 to 10. Students can use preprogrammed spreadsheet software, create their own program, or use paper and pencil computations to find a rating for the song. Inevitably, students are befuddled by the fact that each student's final rating is different. Eventually, someone realizes that the criteria were weighed differently. This is a poignant demonstration of the ways in which merely employing mathematics—a subject often known for its dispassionate accuracy—does not necessarily ensure objectivity. On another level, it is a demonstration of the ways in which mathematics is a fundamentally human enterprise and that without the directed activity of human agents, the mathematics in question would not exist and it seems impossible to separate the objective mathematics from the subjective human elements.

As a next step, students select phenomena that they wish to evaluate using a weighted formula. The activities leading up to this point have helped the students not only know how to construct such a system, but have also helped them recognize the limitations of any such endeavor. There may be instances of irony and contradiction resulting from difficult decisions made along the way. Students are encouraged to choose a topic of interest and to use mathematics as means to think about this interest in a different light.

This lesson, or sequence of lessons, presents a version of mathematics that is different from the typical absolutist version, as the focal point of the lesson is not a transmission of a precast set of truths or skills. But how are the mathematics presented in this lesson different from a constructivist version? The final stage of the lesson makes a fundamental difference quite clear. Once the students have employed the formula to rate a particular instance of some phenomenon, they repeat the process several times evaluating different instances. Students then look at the results and check to see how well the system is working in practice.

Checking the usefulness-in-practice of the music rating formula is somewhat thorny, as the very measures of song quality are problematic in the first place (although we also argue that this is part of the problem's allure). If students created formulae in order to quantify their personal, subjective beliefs about the quality of a song, then they could listen to a number of songs that are unfamiliar to them (say, five) and react to them in a general way, ordering them from favorite to least favorite. Next, they could listen to the songs more carefully, score the songs according to their criteria and scales, and then perform the mathematics required to determine an overall score for each. Comparing their general impressions with the particular scores could be the next step, looking to see whether the formulaic system matched up with the way that they felt about the songs. To the extent that the formula failed to predict whether a song is actually enjoyable, students can troubleshoot. Perhaps errors in calculating or in software programming are to blame. Perhaps the students will decide that they need to tweak the formula in order to change the weight of some criteria, or perhaps this testing has helped the student recognize that an important variable was left out of the equation. A key aspect of this part of the process is that the students, most likely with help and guidance from their teacher, are the ones who decide how to evaluate the quality of their constructions and to test the quality of the evaluation systems that they develop. The introductory chapter of a collection of essays on the importance of modeling in school mathematics places a strong emphasis on this evaluative step: "Modelers must return to the real world during their investigation and compare their mathematical insights and predictions with the actual real-world system. A mismatch … drives the modeler forward, leading to revised choices, assumptions, and decisions" (Cirillo et al., 2016, p. 9). Evaluating one's model is an iterative process, an ongoing toggling between invented and real-world systems. This is likely what distinguishes modeling from the school mathematics many of us remember: "It *begins* and *ends* in the real world" (p. 9).

This lesson grounds the development of new mathematical knowledge, weighted formulae, in a need to solve a genuine problem: to work toward developing a rating system. (In Chapter 8, we explore how evaluating existent weighted formulae is another way to help students gain familiarity with them and to critique their effectiveness in the real world.) Constructing their own models given their individual interests is also important not just as a motivating factor, but also because it shows that mathematics is not simply "stuff" in a textbook or in the head. Instead, it is a practical tool we can use to solve our problems. Finally, testing their formulae in practice provides an important opportunity for students to see and evaluate how well their constructions function. With this lesson, students have not merely been given a tour of Hersh's back part of mathematics, they have been invited to join in the activities taking place there. This mathematizing is at the root of the democratic perspective.

Different Perspectives, Different Implications

It should now be clear how our democratic perspective differs from most constructivist versions of mathematics education. Constructivists tend to view the aim of mathematics education as helping students create their own meanings. While in agreement with constructivists concerning the importance of the recognition of the ways in which students make their own meanings, we claim that the democratic perspective calls for testing individual student beliefs. Quite unlike the absolutist's tendency to test students' beliefs against what they see as true, objective, and external knowledge, a democratic perspective suggests a more functional test. Mathematics ought to be considered in terms of the job that it was created to do. While the absolutist tendency is to hold individual belief up against some measure of external knowledge and constructivists tend to hold up individual belief as all that is possible or desirable, this perspective recognizes that the knowledge that we consider public and tested has come from particular instances of belief that have, over time, shown to work given their intended functions.

Both absolutists and constructivists struggle with philosophical accounts of how mathematics "works." The absolutist has to contend with one of two problems, depending on whether they are Platonist or formalist in orientation. The Platonist cannot adequately explain how we know that we have found answers that correspond to what is existent in the Platonic realm; by definition it is off-limits to human beings. The formalist can demonstrate correctness of an answer according to whether the rules of the tightly bound formal system were followed. However, the formalist has a problem when it comes time to explain how what goes on in the system relates to anything outside of it (such as the physical world and our human activity). Proponents of neither form of absolutism can account for how their version of mathematics—as existing outside of human influence—can be evaluated according to its use. One might be tempted to say that mathematics does not need to be evaluated according to human use, it can simply be evaluated on its own terms. We do not argue this point; however, we do claim that *such a view of mathematics does not go very far toward explaining why it is worth our time to teach and learn it.*

Constructivists run into somewhat similar problems, as the "correctness" of an answer is not judged according to correspondence to an external source, but instead is considered in terms of how it fits into a student's existing mental structures. With inner mental processes as the focal point, constructivists can encounter problems explaining how mathematics means anything outside of individual heads.

The functional account of knowledge suggested by DME judges the "correctness" of a mathematical idea according to how well it helps the holder of the particular belief achieve their desired end. In other words, to the elementary student, 12 divided by 3 is equal to 4 not because it matches some fact in an unknowable realm or because it fits in some way with the inner workings of their mind, but because it helps solve some problem. So 12 divided by 3 is equal to 4 because if the student is in a group of three and has to disperse 12 objects, each student should receive 4. Democratic mathematics education does not stop at this simple empiricism. As students' inquiries become more complex, they will need to employ more complex mathematics. Consequently, the ways in which mathematics will be employed will grow less immediately tied to empirical objects that are able to be physically manipulated.[9]

Ideas regarding the nature of mathematics have real instructional consequences, and the math classroom, in turn, strongly shapes our ideas about mathematics. Though summaries and diagrams sometimes risk essentializing the characteristics of ways of thinking, we feel it is helpful to outline their tendencies and potential educational implications. Hence, Table 7.1 is provided in order to clarify the differing emphases of mathematics instruction according to philosophical orientation.

TABLE 7.1 Summary of Comparison of Philosophies of Mathematics Education

Orientation	Philosophical focal point	Pedagogical focal points
Absolutist perspectives	External focus on structure of the subject matter	Fostering an algorithmic understanding of mathematics Finding the "right" answer
Constructivist perspectives	Focus on internal structures (mental, cultural, social, or material)	Fostering a conceptual understanding (usually of algorithms) Explaining thought process (the right answer is the one that "makes sense")
Democratic perspectives	Focus on functions: mathematics as a tool for human use	Fostering the ability to use mathematics to solve genuine problems Solutions must be tested to see how they function given the purpose and context.

Notes

1 Exactly how such access to the Platonic realm is gained can be problematic. Earlier we explained Kitcher's description of the problems of Platonic intuition or the blending of Platonism with empiricism (see the earlier section in Chapter 3 on Platonism as a philosophy of mathematics, as well as Kitcher, 1983, p. 103).

2 Most reform-oriented mathematics teachers are working to overcome this traditional educational outlook.

3 Toulmin's populational approach is helpful in this regard. In the population of absolutist approaches most will have many commonalities but it is likely that no single characteristic will be exhibited by everyone who can reasonably be considered an absolutist.

4 This constructivist position is still apparent (though far more tempered) in a recent edition of the same text, now authored by Van de Walle, Karp, and Bay-Williams (2019). Under the section "How Do Students Learn Mathematics?" only constructivism and sociocultural theory are mentioned.

5 It should be noted that constructivists of the social sort (e.g., Gergen, Ernest, or the various Marxist theorists) also have concerns about structures, but they are structures pertaining to the culture, social groups, or material conditions. While social constructivism as a philosophy of mathematics certainly does exist, it does not appear to have the same influence on classroom practice, either in degree or extent, as Piagetian structuralism.

6 NCTM published the volume in which Confrey's article appears.

7 A version of this activity first appeared in Warnick and Stemhagen (2007).

8 In non-mathematical terms, this formula represents the belief that lyrics are the most important component of a song. The magnitude of each coefficient shows the weight given to each variable.

9 An extension of Peirce's notion of the role of mathematical models suggests that math students are tasked with creating mathematical models that are more easily manipulated than the empirical objects they represent (Peirce, 1898, pp. 209–216).

References

Cirillo, M., Pelesko, J.A., Felton-Koestler, M.D., & Rubel, L. (2016). Perspectives on modeling in school mathematics. In C.R. Hirsch & A.R. McDuffie (Eds.), *Mathematical modeling and modeling mathematics* (pp. 249–261). Reston, VA: National Council of Teachers of Mathematics.

Confrey, J. (1990). What constructivism implies for teaching. In R. Davis, C. Maher, & N. Noddings (Eds.), *Constructivist views on the teaching and learning of mathematics* (pp. 107–124). Reston, VA: National Council of Teachers of Mathematics.

Goffman, E. (1973). *The presentation of self in everyday life*. Woodstock, NY: Overlook Press.

Hersh, R. (1997). *What is mathematics, really?* New York: Oxford University Press.

Hirsch, E.D. (2016). *Why knowledge matters: Rescuing our children from failed educational theories*. Cambridge, MA: Harvard Education Press.

Kitcher, P. (1983). *The nature of mathematical knowledge*. New York: Oxford University Press.

Millroy, W. (1992). *An ethnographic study of the mathematical beliefs of a group of carpenters*. Reston, VA: National Council of Teachers of Mathematics.

Pajares, M.F. (1992). Teachers' beliefs and educational research: Cleaning up a messy construct. *Review of Educational Research*, 62(3), 307–332.

Peirce, C. (1898). Logic of mathematics in relation to education. *Educational Review*, 209–216.

Prediger, S. (2002). Consensus and coherence in mathematics: Can they be explained in a culturalistic view? *Philosophy of Mathematics Education Journal*, 16. http://socialsciences. exeter.ac.uk/education/research/centres/stem/publications/pmej/pome16/consensus.htm

Rorty, R. (1999). *Philosophy and social hope*. New York: Penguin.

Schoenfeld, A. (1992). Learning to think mathematically: Problem solving, metacognition, and sense making in mathematics. In D.A. Grouws (Ed.), *Handbook of research on mathematics teaching and learning: A project of the National Council of Teachers of Mathematics* (pp. 334–370). New York: Macmillan.

Van de Walle, J. (1990). *Elementary school mathematics: Teaching developmentally*. White Plains, NY: Longman.

Van de Walle, J., Karp, K., & Bay-Williams, J. (2019). *Elementary and middle school mathematics: Teaching developmentally*. New York: Pearson.

Warnick, B. & Stemhagen, K. (2007). Mathematics teachers as moral educators: The implications of conceiving of mathematics as a technology. *Journal of Curriculum Studies*, 39(3), 303–316.

8

MATHEMATICS EDUCATION TODAY

Making Sense of Critical and Democratic Approaches

Thus far we have developed the idea of DME and described its origin story in classically pragmatic terms, with DME arising out of a need created when previous ways of thinking were no longer effective at solving the problems they were designed to address. Of course, intellectual movements are rarely so clean, linear, and directly cause-and-effect. In this chapter we consider another way of thinking about mathematics that has come into being and increased in prominence in recent times. Critical mathematics education (CME) is an approach to school mathematics that, while more focused on practice, also possesses a rich and clearly articulated philosophical base.

Prior to delving into CME a quick note is in order. Over the course of writing this book we have found ourselves most often seeing CME and DME as much more similar than dissimilar. As we read about ideas and activities suggested by CME proponents we frequently see overlap in the two projects. This overlap has also helped us to see some clear differences. As a starting point for this chapter, it is important for the reader to keep in mind that CME and DME are, to our mind, best thought of as largely complementary but with some important differences in focus. Ultimately, we see the development of a robust democratic philosophy of mathematics as a foundational support for those engaging in CME.[1]

Critical Mathematics Education

There are many ways to define and categorize mathematics education that possesses critical dimensions. One core belief common to critical mathematics educators is that constructivist/reform math is not sufficiently focused on meeting the needs of students from marginalized groups or the social justice cause in general. Examples of work that is social justice-oriented and that may or may not be thought of as CME, depending on how it is defined, include efforts to use school math to address equity-equal access and opportunity and some versions of applying culturally relevant pedagogy to math class. These can be viewed as providing important support for broader efforts for social justice through education, and each can be conceptualized as a relatively free-standing category or hierarchically arranged with critical mathematics acting as a foundation for both equity-equal access and culturally relevant pedagogy.

Equity-equal access and opportunity refer to work that seeks to provide a level playing field for all students as well as to work acknowledging that students from traditionally marginalized communities sometimes need more support to achieve in equal measure. Here, mathematics class is seen as a vehicle for increasing opportunity in the world "as it is." Helping students from marginalized groups to perform well on established standardized tests or successfully in college are primary aims in this category. It accepts the curriculum and structure as it is but finds a way to help students succeed in that system. Early iterations of Robert Moses and Charles Cobb's (2001) Algebra Project serve as an excellent example. Upon recognizing that high school Algebra I courses often function as a gateway to college admission, they explicitly designed and implemented a program to help poor, Black students in the rural South succeed in algebra and gain admission to college. This project directly and purposefully uses school math as a lever of opportunity: "(it) is not about simply transferring a body of knowledge to children. It is about using that knowledge as a tool to a much larger end" (Moses & Cobb, 2001, p. 15).[2]

Culturally relevant pedagogy has a history of its own, coming out of the multicultural education and civil rights movements (Banks and Banks, 2007). While not opposed to aims related to traditional success, culturally relevant pedagogy is more concerned with reshaping the educational experience so that it is inclusive and more relevant for those whose needs have not traditionally been sufficiently considered. The multicultural education of Banks and Banks (2007), Gay's culturally responsive teaching (2000), Ladson-Billings's (1995) culturally relevant pedagogy, and Ladson-Billings and Tate's (1995) application of Critical Race Theory to education are several distinct but overlapping versions of work in this vein. In terms of mathematics education, some of this culturally relevant work is also explicitly oriented toward social change (e.g., Gutiérrez, 2013; Gutstein and Peterson, 2013).

Danny Martin is a mathematics educator whose research focuses on the primacy of race in all facets of mathematics education, from the classroom to academic research to educational policy. Extending Ladson-Billings and Tate's work with Critical Race Theory into the realm of mathematics education, Martin (2009) argues that studies addressing issues of race and mathematics achievement tend to rely on uncritical methodologies in which race is grievously undertheorized. Race, he maintains, has typically been used by mathematics education researchers to sift and sort achievement data without attending to sociopolitical considerations of race or to the ways in which children experience racism in the mathematics classroom on a daily basis. Martin instead calls for "research that is grounded in a perspective that conceptualizes mathematics learning and participation as *racialized forms of experience*," hence connecting research to the broader social reality in which math education is embedded (p. 299). In the years since, an increasingly robust Critical Race Theory of mathematics education has helped challenge and reconstruct what we mean by equitable mathematics teaching and learning (see for example Davis & Jett, 2019).

One other strain of culturally relevant mathematics education that merits specific mention is ethnomathematics. Ubiratàn D'Ambrosio locates ethnomathematics at "the borderline between the history of mathematics and cultural anthropology" (D'Ambrosio, 1985, p. 44). Deeply influenced by Lave's social anthropology and the situated theorists (Rogoff & Lave, 1984), there are a number of scholars carrying out research in the area of ethnomathematics theory and practice (e.g., Barton, 1999; Wagner & Borden, 2012; François, Coessens, & Van Bendegem, 2013; Ortiz-Franco, 2013).

The variety of social justice-oriented approaches to mathematics education sketched above can be thought of as existing within or adjacent to CME. Let us now turn specifically to CME. Any account of critical mathematics education needs to start with the Frankfurt School's Marxist theoretical approach of seeking to "critique and subvert domination in all its forms" (Stinson & Wager, 2012, p. 6). Following Stinson and Wager, we echo CME's commitment to "maintain(ing) sociopolitical critiques on social structures, practices, and ideology that systematically mask one-sided accounts of reality which aim to conceal and legitimate unequal power relations" (pp. 6–7). In the US context, CME tends to draw specifically on Paulo Freire's critical pedagogy project. We see some promise for the European version of critical mathematics to aid specifically in efforts toward democratic education, but in what follows we will focus on the Freirean version.[3]

Freire and School Mathematics

Educational researcher Rochelle Gutiérrez's (2013) identification of two foci for critical mathematics educators illustrates the strong connections in CME between mathematics class and social justice and serves as a clear way to begin to consider Freire-inspired CME. She sees CME's two major concerns as: "to (1) develop within learners 'conscientizacao' (a kind of political awareness) that allows an individual to recognize their position in society and as a part of history ... and (2) motivate individuals to action" (p. 41). Paulo Freire was a Brazilian educator who, in an effort to teach poor adult Brazilian farmers to read, came to the realization that simply learning to read was not enough to meaningfully help them improve their lives. Instead, he connected learning to read with learning about the marginalized position in society in which the farmers found themselves and about the power relations that work to sustain it. This experience served as the basis for the philosophy and practice of education as liberation that is most famously explicated in Freire's *Pedagogy of the Oppressed* (1973). In the early 1980s, Marilyn Frankenstein reconceptualized school mathematics using the framework of Freire's critical pedagogy and made a case for the emancipatory dimensions of mathematical and statistical literacy. Her vision of critical mathematics pedagogy is one in which teachers use contemporary issues that are relevant and of interest to students as contexts for learning mathematics skills and concepts. Equally important is that the contexts are designed to "challenge hegemonic ideologies." For example, in a contextual problem related to federal spending, she begins by referencing two articles arguing that the magnitude of the US military budget had been deliberately misrepresented, making it appear much smaller than it actually was. Students are then asked to choose how they would best present the journalists' critique:

> Max and Greenwood's critique of the official statistics on the military portion of the federal budget can be used to learn percents and circle graphs. In addition, students can discuss how they would decide to present the critique, and what aspects of this research and presentation they control. Would they choose to present their critique using raw data, percents, or graphs?
>
> *(1983, p. 330)*

While the journalists' critiques could surely stand alone as "challenges to hegemonic ideologies," students must make key decisions in order to experience the power of (and, we would add, the responsibilities of) statistical representation first hand. They choose what to disclose and what to omit. Here we see intersections between DME and the earliest proposals of CME, not just in the way students take ownership of a project but in the way mathematics is experienced as a tool used with intention. Later in the chapter we revisit this federal-spending problem, picking up where it leaves off here, and describe how DME and CME, at least in this scenario, seem to part ways.

Gutstein's "Reading and Writing the World with Mathematics"

More recently, Eric Gutstein's mathematics education project has become the clearest and most influential work in this vein. We see his *Reading and Writing the World with Mathematics: Toward a Pedagogy of Social Justice* (2006) as providing the core of the contemporary CME movement (at least in the US). His project, that of helping poor, urban youth of color to simultaneously use school mathematics to better understand the sociocultural and political realities that explain their station in life and to employ mathematics to act on and improve this world (all while succeeding in the traditional game of school) cuts across all categories. It is obviously a critical project and it is also just as obviously a culturally relevant way to foster quantitative literacy and to develop some of the requisite skills for democratic participation.

The title of the chapter in which he lays out his theoretical framework makes his commitments clear: "Education for Liberation: Toward a Framework for Teaching Mathematics for Social Justice." He describes his approach as drawing on "Freire's work, and literature on culturally relevant pedagogy and African-American liberatory education" (p. 22). Following Freire, Gutstein centers liberation from oppression in teaching/schooling and regarding mathematics specifically, he explains that "using mathematics—in school, as an educational practice to analyze and affect society—is clearly political and consistent with a Freirian perspective" (p. 23). He draws on Ladson-Billings (1995) to explain that liberatory pedagogy must do three things. It must help students succeed academically, foster cultural competence, and sharpen their understanding of and ability to critique the current world.

The Freirean view of education as an "emancipatory praxis" is in full force in both Gutstein's teaching and scholarship. By emancipatory praxis, Gutstein means "the dialectical interconnection of action and reflection for the purpose of liberation and full humanization" (2016, p. 455). This dialectical movement is essential to Gutstein's liberatory objectives in that it generates the potential for change, often through contradiction, and it is especially apparent in his statement that "we read the world as we shape it and are shaped by it" (p. 456). In terms of approach, Gutstein's book provides a model of his notion of praxis—vignettes of classroom practice and written reflections of his students are interwoven with explication of his philosophical vision of CME. The mathematics class that emerges attempts a balance between mathematics and social issues, usually but not always with the former leading to the latter. Gutstein's first story is about how teaching middle school mathematics on the first anniversary of 9/11 led to much of what became the core of his CME methods. Following a class discussion about 9/11, Gutstein developed an activity related specifically to bombing in Afghanistan. "The Cost of a B-2 Bomber—Where Do Our Tax Dollars Go?" involved scrutiny of US Department of Defense data to discover how much a B-2 bomber cost and then to compare that cost to college education for their school's students. Through this activity the students learned that, for the cost of one bomber, the graduating class from their high school could go to a prestigious out-of-state university for 79 years (2006). The activity included a set of questions in which the students were asked about how this new knowledge made them feel and what, if anything, they think should be done about this situation.

Gutstein lays out his project as primarily involving work toward realizing two kinds of goals: pedagogical and social justice. Gutstein describes them as "dialectically interrelated" and he goes on to explain that "in the larger framework of teaching mathematics for social justice, each set is necessary, and neither is sufficient by itself" (p. 29). Gutstein posits three pedagogical aims, the first of which is reading the mathematical world, or developing mathematical skills and mindsets that increase students' power in the world. The second aim is succeeding academically in the traditional sense. This aim relates to the opportunities afforded by success in school mathematics. Gutstein points out, though, that his is not a "pipeline" argument and that getting more marginalized students through the pipeline to the opportunities that await at the other end without changing the circumstances that created and perpetuated the unequal opportunity is unacceptable. The third aim, changing one's orientation to mathematics, involves an alteration in how students see mathematics. This shift involves a move from mathematics as rote and disconnected to math as a "powerful and relevant tool for understanding complicated, real-world phenomena" (p. 30). In discussing this third aim, Gutstein makes a distinction between the contextual problems of reform math that "often are situated in generic real-world settings" and his CME which more "explicitly ask(s) students to investigate (in)justice" (p. 31).

Again, Gutstein sees these pedagogical aims as dialectically related to the broader social justice aims of CME. He has obviously constructed his pedagogical aims in a way that bends toward social justice. That said, Gutstein is frank about how difficult it can be to have to simultaneously foreground both sets of aims in a public school system not set up for social justice work:

> In Chicago public schools, navigating between, on the one hand, creating problem-posing pedagogies, and on the other hand, ensuring that students acquire the social, cultural, and linguistic capital to pass their courses, have real opportunities, and potentially become agents of social change is extremely complex.
>
> *(p. 31)*

CME and DME: Rivals, Allies or Something Else?

In this section we consider the relationship between CME and DME and how we envision DME as able to support many aspects of the critical project. In order to enrich this discussion, we start with an activity that is an extension of the weighted formula activity presented in Chapter 7 and that can be seen as possessing elements of both CME and DME. The discussion that follows includes a critique of CME and an attempt to reply to the critique, all of which leads to a more thorough consideration of the relationship between CME and DME.

Weighted Formulae: NFL QB Rating Problem

In Chapter 7, we described a lesson in which students develop and test a rudimentary system designed to quantify something fundamentally qualitative: popular music. A next step might be to present a more highly developed (by middle school standards) and respected rating system that utilizes a weighted formula. Using the context of the National Football League's (NFL) quarterback rating system, the activity starts with an informal discussion about what we might need to know in order to judge whether a quarterback is effective, followed by a class debate about the merits of various criteria. Number of completions, touchdowns, wins, sacks, interceptions, rushing yardage, passing yardage, and other statistics are some ideas that usually surface. Interestingly, some of the most insightful ideas come from non-football fans, as their perspective is generally a bit fresher and they are able to think about the situation in a less paradigmatic manner. As a teacher, it does take some skill to engage the non-football fans and to create a discussion climate that feels safe for participation from all. This effort is not only beneficial in terms of the lesson, it is an obligation that classroom teachers face daily.

Next, students are presented with the NFL's formula:[4]

$$R = \frac{50 + 2000\left(\frac{C}{A}\right) + 8000\left(\frac{T}{A}\right) - 10000\left(\frac{I}{A}\right) + 100\left(\frac{Y}{A}\right)}{24}$$

where R = rating, C = completions, A = attempts, T = touchdowns, I = interceptions, and Y = yards gained passing. Discussing the possible reasons why the NFL selected these criteria and not others goes a long way toward demonstrating that the quantitative rating, while informative, is quite limited with regard to the way it represents what actually transpires on the football field. Additionally, although the mathematizing is already well underway at this point, we can begin to focus on the mathematical model, making intuitive sense of the formula. The use of ratios with the number of attempts, (A), serving as the common denominator, the choice of coefficients as a means to "weigh" the various criteria, and the existence of both addition and subtraction (depending on the "positiveness" or "negativeness" of each criterion) are some of the ways students make sense of the formula. Using a spreadsheet program to work with the formula, they experiment with the weighing of criteria (i.e., changing the coefficients in the equation) and with the actual criteria selection in order to evaluate and improve the NFL formula. Students begin to recognize that mathematizing can be a way to make a point or to express themselves.

Students also recognize that the choices involved with mathematical problem-solving are not set in stone and can be challenged. Using mathematics to issue a challenge is a frequent feature in the CME classroom. In our example, students may raise objections to the NFL's official formula even before putting it to the test (why *these* variables and not others?), or they may question its reasonableness after substituting the pertinent statistics for each quarterback to generate an overall rating. The results may be surprising or contradictory, or they may even strike students as suspicious or overtly unfair. Indeed, there has been continued controversy regarding the role of racism and the number of Black quarterbacks in the NFL.[5] The NFL's proprietary quarterback rating formula does not include rushing yards or rushing touchdowns as criteria, two areas in which, historically, many of the best Black quarterbacks have excelled. Here we can see how the weighted formula problem is in some ways consonant with CME, where problems can cultivate sociopolitical awareness, and where doing mathematics stretches beyond executing computational procedures and asks students to analyze, critique, and challenge.

In contrast to CME, or at least in contrast to many of the exemplars featured in CME's growing portfolio of problems and activities,[6] DME is characterized by the way in which philosophical work related to our beliefs about mathematics is brought to the forefront and sustained throughout the arc of the problem. In other words, it is not a post hoc application of mathematical procedures and concepts—learned through some other instructional means—to social justice contexts. As we have said, the DME approach draws attention not just to social realities but to the very nature and limitations of mathematics. In weighted formula problems, our choices of variables and coefficients come with a heavy responsibility, so much so that depending on the context, we may even question the suitability and ethics of mathematically ranking individuals. We may increase our awareness of what mathematics, as a tool in our hands, can and cannot do. In a DME approach, mathematics is therefore not something "taken as it is" but something *in the making*. It is to be read and remade. Doing so is an act that is inherently democratic, strengthening both the enterprise and community of mathematics as well as the democratic capacities and values of the students.

Gaps, Overlaps, and Critiques

Despite their both placing a special emphasis on action, the intellectual tension between democratic/pragmatic and more explicitly critically oriented theorists is long standing. Specifically, critical theorists have tended to view pragmatism's antifoundationalist position as one that precludes the kind of attention to power relations that is essential to emancipatory aims (Kadlec, 2006). Though only obliquely related to this historic impasse, we suggest that DME's focus on learning both the power and limits of mathematics dovetails with the broad CME ethos of challenging power arrangements.[7] Both approaches view mathematics as a means with which to act on the world, and neither approach deifies mathematics. While a more fundamental facet of DME, the idea of recognizing the limits of mathematics can span both CME and DME.

Critical mathematics educators tend to foreground injustice, whereas democratic mathematics educators foreground inquiry and agency. Responding to a manuscript about democratic mathematics pedagogy written by one of the authors, a reviewer wondered whether allowing students to select their own projects would impede the honing of critical tools needed to recognize inequities in society. The reviewer's concern clearly illustrates this difference in focus. Their worry was whether we can trust students to necessarily care about and thus engage in inquiry that will forward the critical agenda. As with CME, we also recognize the importance of the cultivation of interest in fairness and equity among students but we also believe that treating social justice, in some ways, as an indirect aim might be a more effective approach in the long run. Foregrounding inquiry fosters agency *and* ability in students—both invaluable traits for democratic participants. Unfortunately, agency and ability often appear to be a forced choice. The democratic approach seeks to have them reinforce each other and to cultivate them in an intentionally non-dogmatic manner. Entrusting mathematical inquiry—its origin and/or direction—to students brings an earnest significance to what we mean by "student participation." With a democratic approach, problems that are inherently social have no answer key, be they primarily mathematical or moral,[8] but there is a flexible rubric requiring a commitment to self, others, and community. As in the bus stop problem featured at the beginning of Chapter 7, students might wrestle with the consequences of their decisions and have to contend with shortcomings of even the "best" solution. They would take part not just in the production of solutions and models but in judging their rightness and in coming to grips with their righteousness.

The idea of social justice as an indirect aim may seem like something of an affront to CME. However, we propose that prioritizing inquiry is more effective in achieving emancipatory goals in part because it is democratic in its very enactment. In other words, students are already democratic participants rather than preparatory ones. Problems do not merely build agency, they summon it, inviting students to put mathematics and themselves as users of mathematics to the test. In CME, problems are also deeply social in nature and relevant to students' worlds, but they are often designed with particular solutions and conclusions about social and economic disparity already in mind. Considering CME and DME via the quarterback rating activity deepens our understanding of both, specifically showing the many points of commonality between the two approaches as well as some points of departure. Next, we draw out two specific points of departure requiring further critical analysis.

CME and the Location of Agency

The tendency to be prescriptive forms the basis of our first critique of CME. DME is less direct and less certain but ultimately more likely, by virtue of students practicing democratic habits, to leave them wanting to be a part of a more just world. Drawing on Dewey's "more verb than noun" concept of democracy, Stitzlein argues for the alignment of aim and enactment:

> [Democracy] is a way of life that we strive to achieve, but in order to do so, our day-to-day practices must themselves also be democratic. This includes the way that children are educated to be good citizens. We cannot indoctrinate them into seeing democracy as an admirable end goal while engaging in classroom practices that are totalitarian. Rather, we must employ means that are aligned with the end …
>
> *(2014, p. 62)*

While "totalitarian" is surely an unfair characterization of any aspect of CME, there can be something of the indoctrinaire in CME's content and delivery. As an example, we return to Marilyn Frankenstein's proposed problem about the US military budget. After students negotiate ways to present the journalists' critique, Frankenstein suggests pursuing the following question and discussion topic:

> Do [students] agree with Max that the space program should be considered part of the cost of "Past, Present and Future Wars"? Discussing how to present the statistics to demonstrate that the United States is a welfare state for the rich can include practice of arithmetical operations.
>
> *(p. 330)*

While it is possible that students might disagree with the journalist's proposal about reclassifying the space program, murmurs of disagreement are simply not under consideration in the discussion activity in which students negotiate how, but not what, to present. "The United States is a welfare state for the rich" is a strong and thought-provoking statement, one supported by what we suspect is keenly perceptive journalism, but it is a conclusion reached *for*, not *by*, students. Arguably, it is the teacher, not the students, whose agency is primarily established and affirmed through this activity.

CME's direct and unambiguous messages about injustice are understandable, given the urgent need for swift and certain social change. The very recognition of the capacity of school mathematics to help realize change is one of its most important and enduring contributions. Indeed, both CME and DME recognize that the frequent omission of math class from civics education is a missed opportunity in a time when we seem to need it more than ever. CME proponents may thus view the uncertainties of DME in practice (e.g., whether students choose to pursue questions of social import, or whether they arrive at solutions that are morally defensible) as a risk not worth taking. In our view, conversely, it is a necessary fusion of means and ends.

CME and Application over Philosophy

In his genealogy of mathematics education, cultural and educational theorist Houman Harouni positions CME as a project that uses "social-analytical mathematics" for its emancipatory purposes. In a modern (and admittedly loose) sense, social-analytical mathematics means data analysis, a form of applied mathematics replete with human intention, though any intention or bias may go unnoticed by the consumer. It is this type of mathematical knowledge that Frankenstein argues "can be used to obscure economic and social realities" and therefore must be read with a degree of skepticism (1983, p. 315). That which is "to be read" comes from outside the classroom walls, as do the techniques used to decode and critique it:

> [Frankenstein] treats the basic materials of math, both in numerical and technical terms, as having already been provided by another source, one that

is essentially suspect. In fact, she relies on an unnamed and undescribed form of training that is supposed to equip students with the basic technical skills that enable them to analyze data, which is also, more often than not, gathered elsewhere. This outsourcing of basic technical training is endemic to critical math education.

(Harouni, 2015, p. 68)

The outsourcing of mathematics, or what we have described as the manner in which CME largely leaves math "as it is," is at the heart of our second critique. The nature of mathematics is never brought into question: mathematics just *is*. As stated earlier, CME in practice is typically an application of mathematics to social justice contexts. The learning of mathematics—its requisite skills and concepts—is usually relegated to whatever curriculum is in place, be it traditional, reform, etc. The "stuff" of mathematics (e.g., proportional reasoning, measures of central tendency) is determined elsewhere by school districts, state, and/or local standards, or even by curriculum publishers. Beliefs and assumptions about mathematics are rarely brought to the forefront. DME, however, works to link how we think about math with its teaching and learning instead of applying it unquestioningly. Students are brought along in the democratic effort to continually reconstruct the enterprise.

One reasonable response to this critique could be that Gutstein's focus is, in large part, the result of his recognition of what his students need. In a sense he is trying to do the work of the equity-minded mathematics educators mentioned earlier (e.g., Moses and Cobb) while also providing a liberatory experience for his students. Gutstein's version of CME involves a series of difficult decisions about how to allocate precious class time, and his solution seems to include teaching the existing math in order to ensure his students get the opportunities that school math success can provide while preserving enough time to apply the math in ways that will raise students' critical consciousness.

Andrew Brantlinger, a critically oriented mathematics educator, provides an excellent description of how tensions between these twin aims played out in his efforts to teach math in a way that might accomplish both (2013). He set out to put a critical approach into practice in the geometry class he taught at an urban alternative high school. In the end, Brantlinger found it difficult to teach both critical and more mainstream mathematics for a variety of reasons, ranging from the relative incompatibility of geometry and social issues, to the resistance of his students to recognize how critical mathematics could benefit their lives. Some of his students also worried that the critical focus would get in the way of their learning math as a means of positioning for college. Brantlinger skillfully describes the difficulties he encountered in designing and carrying out his project. He deeply questions his practices and even his underlying aims, all in the name of improving—rather than abandoning—critical mathematics education.

Whether math teachers implement a traditional curriculum or, like Brantlinger, a reform curriculum faithful to a constructivist approach, synthesizing content and critical math objectives is an extraordinarily difficult task. As Brantlinger discovered, the objectives at times seem quite disparate. Our hope is that DME can make the critical math project more doable and bypass the frustrations that come from trying to satisfy what appear to be two competing aims. In other words, while CME proponents see their approach as a way to be able to contend with the content–justice dichotomy, the dichotomy is built into CME. The DME focus on the remaking of mathematics serves to blur the sharp lines between content and justice that cause such trouble for those working to implement CME in their classrooms.

While a clear presence in CME, and though it may seem incongruous, the intrinsic division between content and justice is also detectable in the work of some researchers who share our interest in school mathematics and democratic aims. For example, Annica Andersson and Lisa Österling set out to explore the relationship between democratic actions (e.g., equitable participation and collaboration, students shaping the direction of math activities) and what students actually valued in mathematics class. The results pointed to a disappointing disharmony between democratic actions and student values. Despite a strong cultural and official endorsement of democratic actions in the Swedish mathematics curriculum, students valued items such as "explaining by the teacher" and "knowing the times tables" over more patently democratic choices such as "discussion" and "students posing maths problems" (2019, pp. 80–81). Even echoes of absolutism reverberated among the researchers' findings in that "students value control through certainty and the mastery of rules" (p. 85).

A more careful reading of this and other publications featuring the term "democracy" and its variants confirms what we see as both the convergences with and divergences from our vision of DME.[9] A key figure in CME and an outspoken proponent of democracy in school mathematics, Ole Skovsmose, has called for the need for CME to resist dogmatism and to remain open to change:

> one cannot assume any specific interpretation of social justice, mathemacy, dialogue, etc. They are all contested concepts. They are under construction. The open nature of critical mathematics education is further emphasised by the fact that forms of exploitations, suppressions, environmental problems, critical situations in general are continuously changing.
>
> *(Skovsmose, 2016, p. 12)*

Skovsmose's prescription is for critically-oriented mathematics proponents to embrace uncertainty: "Critique cannot develop according to any pre-set programme. As a consequence, the basic epistemic condition for a critical activity is uncertainty" (p. 12).

We heartily agree with Skovsmose's call for the need for CME proponents to remain open and to resist static definitions of its key concepts. Regarding Skovsmose's overlap with DME, while the word "democratic" features prominently in much of his work, his version of democracy seems to include this infusion of openness or uncertainty in nearly every facet except mathematics. The conspicuous absence of mathematics itself as something "under construction" and thus as a field of democratic possibility is evident in Skovsmose's quotation about the many contestable facets related to mathematics education. We are heartened by the humility evident in Skovsmose's approach and see it as an important development in CME. For this humility to be even more useful to CME, it needs also to be applied to CME adherents' aims, as dogmatically adopting particular ends can get in the way of open democratic learning experiences for students. For Skovsmose's humility to be of full use, however, it requires overcoming this sharp separation between content and purpose and hence, between mathematics and morality.

DME and CME: Critique, Dialogue, and Hope

The writing of this book coincided with the long overdue racial reckoning that intensified in the summer of 2020. Our argument in favor of the indirect aspects of DME was not resonating with the way that the immediacy of the demonstrations and protests for abolition of racist structures was providing momentum needed to make progress toward overcoming the weight of history and tradition. While no less convinced of the need for urgent immediate action, we also have come to see new possibilities as to what the deliberative and dialogical aspects of DME have to offer. Our claim is that DME's linking of the remaking of mathematics by students and their use of mathematics to understand and change our social world are key ways that mathematics education can aid in the schools' effort to foster fully engaged democratic citizens. Applying math or other critical tools that already exist to today's problems can help raise awareness and possibly even a desire to change inequitable and oppressive structures, but it does not leave sufficient space for hope as we think is necessary to realize the democratic aims of a just and equitable society. As Stitzleen explains:

> [Hope] entails action and effort even in the face of current limitations, and a confidence that they can be overcome. In this way, we see why it is worthwhile to understand hope as a habit that entails action—especially actions that engage proclivities toward change …
>
> *(2014, p. 77)*

Many educators have heard the call of the racial reckoning and have started by looking anew at their content—looking to see whose voices are represented, where the sources of knowledge come from and why, and taking steps to ensure that the story of their discipline is told in a way that enhances diversity and is actively antiracist. As we have lamented, the content of math class is all too often seen as existing outside of social and political contexts. We see great possibility in

DME to serve as a foundation for a way of thinking about the content of mathematics class that will make it impossible for it to be thought of as impervious and unanswerable to social movements. It is all too easy to imagine curricula in all other content areas changing to meet our evolving standards of equity and math getting left behind. Again, the philosophical core of DME—its attentiveness to contingency and our human contexts and problems—is a necessary precondition for the adoption of a school mathematics that can make it vital in the effort for racial justice.

It is also important not to conflate the relative "indirectness" of DME with an incremental approach to social change in wider society. We see DME as fostering the kind of democratic participants who have the skills, knowledge, and desire to take urgent and immediate social action. While CME is designed to hone students' critical edge and presumably their ability to participate in or even create movements and the altering of structures that lead to urgent social change beyond the walls of the school, it is not evident that CME is better able to do this than DME. We now turn to philosopher of education Aaron Schutz's critique of Deweyan democratic approaches, as it also seems to apply to critical approaches. Dewey believed that because schools in a democracy were not supposed to merely seek to reproduce existing culture, they needed to teach students how to change society. According to Schutz:

> Dewey's point, then, was not that what happens in the school must be the same as what happens in the outside world, but instead that what students learn in schools must be useful in the activities they will engage in when they leave the school.
>
> *(2001, p. 279)*

Schutz uses Mayhew and Edwards' admission that some former Laboratory School students found that the world was not so easily changed to argue that the school was not successful in creating democratic participants, and that it "failed to achieve Dewey's most fundamental aims for education" (p. 279). This brings to light a paradox of Deweyan democratic education: a school is part of society and a mini-democracy, but at the same time it is also a distinct social organ. We see benefits and drawbacks to conceiving of school in both ways. Exclusively learning how to fit into our existing society only guarantees stasis. At the same time, creating a barrier between the messiness and disappointments of our current society and the school in order to buoy students' idealism and resistance to the status quo seems similarly fated. It is at least as likely to lead to an inability of students to engage with and change the world. There is no easy way around this paradox, but we do not see Schutz's critique as fatal to the idea of the Deweyan approach to democratic education, nor do we see the critical approach as free from the need to contend with this paradox. In the end, we believe that because DME practitioners ask students to act as moral and mathematical agents simultaneously, and that because DME places trust in students to launch inquiry and change mathematics itself when warranted, it is more likely to cultivate the veritable change agents we so desperately, and urgently, need.

Critical approaches are quite effective at critique but often not so effective at charting new courses of action.[10] Interestingly, democratic approaches, at least Deweyan ones, have been described as useful when everyone is ready to come to the table, but not so useful when a community's various factions are not interested in or ready to come together. Perhaps CME and DME are not just related; one could argue that they need each other. That said, we see the differences between the two approaches as important. While we enthusiastically agree with CME's mission to see school mathematics as a place to go beyond rote learning and to engage students as civic participants, we worry that CME is only critical on one side of their content/justice divide. Mathematics content is largely taken as it is ("outsourced"), and even on the justice side, the assumptions about criticality are often handed more or less premade to the students from the teacher. With DME, even when thinking of math class through CME's dichotomous content–justice lens, both the math content and its application to social issues are not beyond critique. Furthermore, DME's emphasis on the joint/communal aspects of doing mathematics fosters the responsibility to others that can often be missing from more individually oriented approaches to teaching and learning.

We see DME as a way to aid in the dialectical relationship between Gutstein's pedagogical and social justice goals. Furthermore, DME's explicit philosophical work helps transform school mathematics into something far more than a post hoc application of the subject matter to unjust social situations. The choice to prioritize students' questioning and reconstruction of mathematics makes it more likely that students will be empowered and interested in using mathematics for social justice efforts.

A final note: we hope that our relationship to CME isn't too puzzling. One way to look at it is to place it in the larger context of this overall project. Looking back to the math wars that pit traditionalist against constructivist math educators, we see ourselves as mostly sympathetic to constructivism. In fact, we see ourselves more or less as offering a certain version of constructivism. This has generated some seemingly strange points of congruence and conflict between constructivism and DME. For example, DME's focus on student agency and (re)construction sits directly in mainstream constructivism, though constructivism's goal of building conceptual understanding of school math "as it is" directly contradicts the DME ethos. Likewise with DME and CME, we see our ideas as largely fitting into the CME thought world but, as noted in this chapter, with certain important differences. Our critique is rooted in a desire to find better ways to forward many key parts of the CME agenda. We see the development of DME as a necessary next step in the journey toward a school mathematics that will help enact the broader democratic visions shared by both CME and DME.

Notes

1　See Stemhagen (2016) for an earlier effort to consider the relationship between critical and democratic mathematics education.
2　While the Algebra Project is an example of an effort that seeks to address inequitable access to high quality school mathematics opportunities, it should also be noted that with its roots in Moses's involvement in the civil rights movement, it also puts school math class in its wider social and political contexts. The Algebra Project has influenced the creation of a variety of organizations that expressly link math and social change. See, for example, the Baltimore Algebra Project's website: https://www.410ap.org
3　An interesting development of late has been the increasing use of the term "democracy" by mathematics educators. For example, Ole Skovsmose (whom Ernest has dubbed the father of CME) has turned toward democratic language in recent works. While this could be seen as supporting our assertion that DME and CME overlap in many ways, we also see the use of Skovsmose's democratic language as largely existing within the CME frame and not necessarily signaling a substantive move toward a robust Deweyan democratic approach. Later in this chapter we explore these questions in more detail.
4　This is only an approximation of the formula, as the NFL does not make their formula public. Football fans and statisticians have developed many variants of the formula by working backward from the finished quarterback ratings and the statistics that are used to judge.
5　For example, see the blog post by Peter Keating (2010): "Black QBs are underrated and underpaid."
6　By this, we mean activities described in CME scholarship as well as in resources dedicated specifically to teaching math for social justice, most notably Gutstein and Peterson's "Rethinking Mathematics: Teaching Social Justice by the Numbers" (2013).
7　The weighted formula activity and specifically the ways in which a quantified rating system for songs stretches the usefulness of math, probably past the breaking point.
8　Within DME there is no sharp distinction between mathematics and morality, as mathematics is inherently social and linked to human needs and aims.
9　Democracy has been invoked by more mainstream reformers, too. While we are excited by any efforts to link math class to our broader social/political realities, reform versions of democratic mathematics tend to focus more on mathematical skill development and understanding as opposed to agency, responsibility, empowerment, etc. For an example of this reform version of democracy and mathematics, see Steen's (2001) *Mathematics and Democracy: The Case for Quantitative Literacy*. There are other mainstream mathematics educators who invoke democracy in other interesting ways. See Philip and Rubel's (2019) *Classrooms as Laboratories of Democracy: The Role of New Quantitative Literacies for Social Transformation*. In it they add cooperation and community to their socially conscious application of data literacy.
10　The abolition movement is one possible exception to this. See, for example, Allegra McLeod's "Envisioning Abolition Democracy," in which she provides critique of existing systems/structures as well as some very clear prescriptions for how to move toward justice.

References

Andersson, A., & Österling, L. (2019). Democratic actions in school mathematics and the dilemma of conflicting values. In P. Clarkson, W.T. Seah, & J. Pang (Eds.), *Values and valuing in mathematics education: Scanning and scoping the territory* (pp. 69–88). New York: Springer Open.

Banks, J.A., & Banks, C.A.M. (2007). *Multicultural education: Issues and perspectives*. Chichester and New York: Wiley.

Barton, B. (1999). Ethnomathematics and philosophy. *Zentralblatt für Didaktik der Mathematik*, 31, 54–58.

Brantlinger, A. (2013). Between politics and equations: Teaching critical mathematics in a remedial secondary classroom. *American Educational Research Journal*, 50(5), 1050–1080.

D'Ambrosio, U. (1985). Ethnomathematics and its place in the history and pedagogy of mathematics. *For the Learning of Mathematics*, 5(1), 44–48.

Davis, J., & Jett, C. (Eds.) (2019). *Critical race theory in mathematics education*. New York: Routledge.

François, K., Coessens, K., & Van Bendegem, J-P. (2013). The spaces of mathematics: Dynamic encounters between local and universal. In P. Smeyers & M. Depaepe (Eds.), *Educational research: The importance and effects of institutional spaces* (pp. 135–152). New York: Springer.

Frankenstein, M. (1983). Critical mathematics education: An application of Paulo Freire's epistemology. *Journal of Education*, 165(4), 315–339.

Gay, G. (2000). *Culturally responsive teaching: Theory, research, and practice*. New York: Teachers College Press.

Gutiérrez, R. (2013). The sociopolitical turn in mathematics education. *Journal for Research in Mathematics Education*, 44(1), 37–68.

Gutstein, E. (2006). *Reading and writing the world with mathematics: Toward a pedagogy for social justice*. London and New York: Routledge.

Gutstein, E., & Peterson, B. (2013). *Rethinking mathematics: Teaching social justice by the numbers* (2nd ed.). Milwaukee, WI: Rethinking Schools Ltd.

Harouni, H. (2015). Toward a political economy of mathematics education. *Harvard Educational Review*, 85(1), 50–74.

Kadlec, A. (2006). Reconstructing Dewey: The philosophy of critical pragmatism. *Polity*, 38(4), 519–542.

Keating, P. (2010, November 4). Black QBs are underrated and underpaid. *ESPN*. http://m.espn.com/general/blogs/blogpost?blogname=keating_peter&id=5764668&category=Sport~90~90&wjb=

Ladson-Billings, G. (1995). Toward a theory of culturally relevant pedagogy. *American Educational Research Journal*, 32, 465–491.

Ladson-Billings, G., & Tate, W.F., IV (1995). Toward a critical race theory of education. *Teachers College Record*, 97(1), 47–68.

Martin, D. (2009). Researching race in mathematics education. *Teachers College Record*, 111 (2), 295–338.

McLeod, A. (2019). Envisioning abolition democracy. *Harvard Law Review*, 132(6), 1613–1649.

Moses, R.P., & Cobb, C. (2001). *Radical equations: Civil rights from Mississippi to the Algebra Project*. Boston: Beacon Press.

Ortiz-Franco, L. (2013). Chicanos have math in their blood. In E. Gutstein & B. Peterson (Eds.), *Rethinking mathematics: Teaching social justice by the numbers* (2nd ed.) (pp. 95–111). Milwaukee, WI: Rethinking Schools Ltd.

Philip, T., & Rubel, L. (2019). Classrooms as laboratories of democracy: The role of new quantitative literacies for social transformation. In L. Tunstall, G. Karaali, & V. Piercey (Eds.), *Shifting contexts, stable core: Advancing quantitative literacy in higher education* (pp. 215–223). Washington, D.C.: MAA Press.

Rogoff, B., & Lave, J. (1984). *Everyday cognition: Its development in social context*. Cambridge, MA: Harvard University Press.

Schutz, A. (2001). John Dewey's conundrum: Can democratic schools empower? *Teachers College Record*, 103, 267–302.

Skovsmose, O. (2016). Critical math education: Concerns, notions and future. In P. Ernest, O. Skovsmose, J.P. van Bendegem, M. Bicudo, R. Miarka, L. Kvasz, & R. Moeller (Eds.), *The philosophy of mathematics education* (pp. 9–13). New York: Springer Open.

Steen, L.A. (Ed.) (2001). *Mathematics and democracy: The case for quantitative literacy*. National Council on Education and the Disciplines. Princeton, NJ: Woodrow Wilson Foundation.

Stemhagen, K. (2016). Deweyan democratic agency and school math: Beyond constructivism and critique. *Educational Theory*, 66(1–2), 95–110.

Stinson, A.A., & Wager, D.W. (2012). A Sojourn into empowering uncertainties of teaching and learning mathematics for social change. In D.W. Wager & A.A. Stinson (Eds.), *Teaching mathematics for social justice: Conversations with educators* (pp. 3–18). Reston, VA: National Council of Teachers of Mathematics.

Stitzlein, S.M. (2014). Habits of democracy: A Deweyan approach to citizenship education in America today. *Education and Culture*, 30(2), 61–86.

Wagner, D., & Borden, L.L. (2012). Aiming for equity in ethnomathematics research. In B. Herbel-Eisenmann, J. Choppin, D. Wagner, & D. Pimm (Eds.), *Equity in discourse for mathematics education: Theories, practices and policies* (pp. 69–88). New York: Springer.

9

DEMOCRACY, MATHEMATICS, AND EDUCATION REVISITED

School mathematics has the potential to aid in the wider civic and social purposes of school, and seeking to accomplish these aims can, we believe, lead to better mathematics teaching and learning. While the book has traversed between giving attention to underlying ideas about the nature of mathematics and also to math pedagogy, this project has all been in the name of developing a philosophy of mathematics that will make democratic teaching and learning more likely. The meaning of "student empowerment" varies according to different mathematics educational approaches, and it has a particular essence in the context of DME that helps make clear the connection between math and morality. We conclude with some thoughts about how this democratic philosophy of mathematics can support various curricular, pedagogical, and even policy possibilities.

Empowerment as an Aim of Mathematics Education

Early on, we claimed that differing aims of mathematics education account for some of the intractability of the math wars. While we would never argue that teachers see mathematics as a means of actively disempowering students, we do believe that, unfortunately, mathematics classes frequently have such an effect. Our claim here is that if empowering students is an aim of mathematics education (and we argue that it ought to be if increased social equity and democratic participation are more general aims of education), then adopting a democratic perspective of mathematics can help move toward this aim.

The absolutist's portrayal of mathematics as something above, beyond, and somehow better than human knowledge sends an initial message of the fundamental unworthiness of each of us to really experience mathematics. Furthermore, by presenting mathematics as a static body of external (both to us and the physical world) knowledge, it is difficult to teach mathematics in a manner which shows its usefulness. Constructivists' preoccupation with the ways in which classroom mathematics meshes with students' existing mental structures can be viewed as an overcorrection of sorts in that students' thinking becomes the focus, to the exclusion of its external relevance and soundness. Students' mathematical constructions are held up as unique and valuable in their own right and at times students do not recognize that much of mathematics' power lies in its usefulness. In other words, constructivism can send the false message that all mental constructions of mathematics are equally good and equally practical. The democratic perspective, undergirded by an evolutionary view of mathematics, encourages teachers to advance an understanding of mathematics that encourages students to actively construct mathematical ideas but also to test them to see how well they work, given their purposes and the context in which the work is undertaken.

Regarding empowerment and mathematics education, we recognize how contrary it may seem to critique the view that students' constructions are meritorious in and of themselves. After all, wouldn't valuing such constructions serve only to empower students and fortify their sense of mathematical self-efficacy? Paul Ernest, whose particular brand of social constructivism we detailed in Chapter 4, would surely say that it would, asserting that when students view themselves as "knowledge sources," they are epistemologically empowered (Ernest, 2002). By contrast, DME inherently works toward the development of epistemological empowerment by emphasizing the ways in which we create mathematical knowledge in order to live in the world. Instead of turning inward,[1] as many constructivists tend to do, we argue that a more effective route to epistemological empowerment is to turn toward human activity in the environment. Indeed, DME requires teaching mathematics in a way that helps students see the contents of the discipline as not coming ready-made from outside of the human realm but as evolving over time as a means to further human inquiry. Furthermore, DME requires that students learn to engage in mathematical practices, mathematizing as a means to further their own inquiry and to solve the real problems that surround them.

Ernest speaks of another type of empowerment, social empowerment, that comes from using mathematics to "better one's life chances" (2002, p. 1). He explains that the world in which we live is highly quantified and hence already mathematized. From wage structures to train schedules to credit scores, the mathematical machinery of our everyday lives is ubiquitous yet often invisible to us. These externally imposed mathematizations are, to Ernest, necessary to understand and critique in order to ably negotiate the world:

> Unless schooling helps learners to develop the knowledge and understanding to identify these mathematisations of our world, and the confidence to question and critique them, they cannot be in full control of their own lives, nor can they become properly informed and participating citizens.
>
> *(p. 7)*

Ernest's linking of mathematical understanding to social empowerment bears a resemblance to Marilyn Frankenstein's argument that mathematical and statistical literacy are, in a sense, passkeys to emancipation. Educating for increased technological knowledge, as she calls it, can work to render the invisible visible and expose the ways in which this mathematical machinery can "obscure economic and social realities" (1983, p. 315). In so doing, Frankenstein lays bare the non-neutrality of mathematics and its applications.

Elsewhere, the case made for quantitative literacy has had less to do with mathematics' non-neutrality than with the increasing need for a numerate citizenry (where "numeracy" is the mathematical analogue of literacy, in the traditional sense) in our data-deluged times. According to this line of thinking, as the world becomes increasingly quantitative, we must either keep up or risk losing our democratic way of life altogether. In the preface to a collection of essays on mathematics, democracy, and quantitative literacy, Robert Orrill chillingly describes the potential fallout from an innumerate populace: "If individuals lack the ability to think numerically they cannot participate fully in civic life, thereby bringing into question the very basis of government of, by, and for the people" (Orrill, 2001, p. xvi). Here, quantitative literacy is recognized not for its emancipatory benefits but for its crucial role in preserving the foundations of self-rule. But Orrill also says that a world bursting with math is still a world where, with the proper tools in hand, a person can be an effective change agent: "Potentially, if put to good use, this unprecedented access to numerical information promises to place more power in the hands of individuals and serve as a stimulus to democratic discourse and civic decision making" (p. xvi). There are indeed traces of Ernest's "properly informed and participating citizens" but, notably, none of Frankenstein's watchfulness. The integrity of mathematics is not under scrutiny here, only a citizen's ability to comprehend it.

Orrill's qualifier, "if put to good use," is a loaded reminder that the social and environmental repercussions of a technological age, be they beneficial or deleterious, are not wrought by mathematics but by its administrators. Mathematics is a tool to be used with responsibility, and no one more convincingly tells the story of what comes from math applied irresponsibly than Cathy O'Neil. In her best-selling 2017 book *Weapons of Math Destruction*, O'Neil exposes how the models and algorithms that have a bearing on our lives are frequently left unchecked and unchallenged, as if they are somehow objective and fair simply by virtue of their being mathematical. In her example of a weighted formula gone terribly wrong, she tells the story of how Washington, D.C. Public Schools partnered with external consultants (presumably experts in statistical analysis) to develop a scoring system for evaluating teacher performance. Based on relatively few factors, most prominently students' standardized test scores, the algorithm was not subjected to any rigorous evaluative process—in part because doing so would have required years of data accrual and a vastly larger sample size. Not only was the algorithm's integrity unquestioned by those in power, but its very design was not disclosed to teachers, who thus had no knowledge of how they would be evaluated. In the end, administrators used the resulting scores to justify firing teachers, no matter how respected or beloved. The algorithm had the ability to, and in fact did, "turn someone's life upside down" (p. 10), and the health of the community was adversely impacted as well. "Weapons of math destruction" such as the D.C. teacher evaluation model are opaque to everyone but those who create them, and those who create them are heedless of the power and limitations of the tools they wield.

O'Neil's message about the dangers of mathematics used indiscriminately has a place in our conception of empowerment as an aim of school mathematics, but it also has a place in moral development as well. In DME, students are empowered by pursuing their own paths of inquiry and using mathematics to act meaningfully in and on the world. They subject their models to testing and retooling by putting results into immediate context. In effect, they assess the model's social utility and goodness. But the role of decision maker, no matter how empowering, comes with its own set of troubles. Agency means taking action, and, as Glaude has said, "The world of action … is fraught with uncertainty" (2007, p. 22). Mathematics and morality constitute a tightly woven fabric. Math cannot on its own, no matter how much confidence we have in its stability and internal integrity, promise social benefit.

The Relationship between Mathematics and Human Values

We turn now to another scene, this time involving a kindergarten classroom in Japan. The children are working on a contextual problem in which they serve a total of five sweet potatoes to two distinct groups of animals, a group of mice and a group of moles. Using actual sweet potatoes as manipulatives, they must determine how many they would give to each group. The mathematical objectives of the activity include building flexibility in decomposing (breaking apart) whole numbers into two addends. In other words, the children could give three potatoes to the mice and two to the moles, or they could give two to the mice and three to the moles. Elementary math teachers know very well how rich these activities can be and listen attentively for emerging mathematical ideas. For example, children may use everyday language to describe the commutative property of addition when noticing that "two and three" is the same total amount as "three and two." There may even be evidence of early algebraic reasoning when children compare the combination of two and three to the combination of one and four, noticing that the total amount, five, remains unchanged because "one moved." But the focus of Nagisa Nakawa's 2019 study, from which this vignette is taken, is not properties of operations or early algebraic reasoning. Rather, it is the manner in which the children's personal and social values propel their mathematical values. From our perspective, social and mathematical values have the capacity to propel and enrich one another.

As the lesson in Nakawa's study unfolded, the children demonstrated that they were deeply concerned about sharing five potatoes fairly between two groups. Might they cut a potato in half? To help keep the focus trained on whole number combinations of five, the teacher explained that "we do not have any knives to cut" (p. 162). Given this constraint, the children looked to other, more complex variables in order to contend with their sense of fairness and the numbers at hand. The size of the actual sweet potatoes mattered to some children, such that giving three small sweet potatoes to the mice and two large sweet potatoes to the moles was justifiable. Others considered the size of the animals, or even the relative size of the animals' mouths in their decision making. Because the mathematical objective was to generate as many different combinations as possible, one pair of kindergarteners came up with "five and zero," which greatly troubled their classmates. Nakawa notes how, in general, the children were "more attentive to fairness than to new mathematical findings" (p. 164). Still, it is important not to misinterpret this moment as mathematics taking a back seat, so to speak, to social values. It instead marks a fusion of mathematics and morals, where concepts of equivalence, and even of the number zero, are forged through a context of caring.

Both Nakawa's study and the research of Andersson and Österling (introduced in Chapter 8) appear in a 2019 compilation of essays about "values and valuing" in mathematics education, a compilation intended to disrupt the myth of mathematics as a value-free enterprise. In comparing these two studies, we were struck by how the deeply embedded assumptions about the nature of school math seemed at odds with one another. In Andersson and Österling's study, the content–justice dichotomy is embedded in the researchers' assumptions that democratic values are endemic to various aspects of classroom life, but not to the subject matter at hand. In Nakawa's study, however, mathematical and moral development are tightly interwoven. The words and actions of the kindergarteners demonstrate not only how mathematics was and is born of problematic situations but also that our solutions are mediated by our humanity. Math and morality are organically joined, as young children readily show us without any prompting. Reading Nakawa's and Andersson and Österling's reports in succession tells us something important: by eschewing the philosophical work of probing and questioning the nature of mathematics, we risk breaking the bond between content and justice, between math and morality. Unfortunately, in school mathematics these bonds are often not even able to be recognized, as mathematics is seen virtually by definition as unrelated to our human pursuits, interests and most damagingly, our values. In effect, the math–morality connection is schooled out of them. DME's blurring of the content–justice dichotomy starts as an abstract philosophical exercise but is designed to help make the practices of school math more able to reconcile this math–values distinction.

Final Thoughts on the Role of Philosophy in Mathematics Education

A side benefit of this project is that it makes a case for the worth of thinking deeply about mathematics, its teaching and learning, and the wider contexts within which all of it sits. In other words, the book in its entirety serves as an example of why the upstart field of philosophy of mathematics education matters. The humanities are under pressure in general. In social science-influenced institutions such as schools of education, the humanities have been shrinking for decades, often pushed aside in favor of empirical (usually quantitative) and more directly practice-oriented approaches. This effect is magnified in the realm of mathematics education, as the subject matter, to many, seems completely unrelated to the human part of the humanities. What we have attempted with this project is to engage in a humanities-oriented enterprise, primarily a philosophical one, in order to provide ways of thinking that will aid in the practice of mathematics education and in efforts to create a more fair, equitable, and participatory democracy.

Taking a humanities-based approach to the problems of mathematics education allows us to see things from a different perspective. We are freed from the need to remain tethered to a notion of "success" that is tightly linked to performance on high stakes tests or the accumulation of completed math courses. Instead, we are able to look at school mathematics in its wider contexts. In this project, we were able to employ historical and philosophical lenses that enabled us to see the problems of mathematics education in a different light. Framing contemporary mathematics education's issues and problems as rooted in a series of philosophical disputes opened up potential solutions that would likely otherwise never have surfaced. DME rose somewhat gradually, starting with recognition of the philosophical underpinnings of the math wars—the conflict between proponents of traditional and constructivist approaches to math teaching and learning. We developed a pragmatic/evolutionary philosophy of mathematics as a way to get beyond the problems that accompanied the forced choice of mathematical absolutism or constructivism that undergirds the math wars. This philosophy is designed to transcend the historical "algorithms vs. understanding" tension by focusing on local, relevant human contexts and the purposes of mathematical activity. The second part of DME came about as a way to contend with what we saw as an emerging problem in mathematics education: the conflict between mainstream mathematics educators and CME adherents. The resulting DME involved fusing the pragmatic/evolutionary perspective with the way that mathematics fits into broader social and civic realms. In other words, it linked Dewey's philosophy of mathematics education with his political philosophy.

The fledgling field of philosophy of mathematics education helped make the humanities-oriented approach of this project possible. Mainstream accounts of the subject matter of this book would likely tend to silo mathematics, philosophy of mathematics, and philosophy of education. Philosophy of mathematics education as an integrated area of study makes it more possible to blur the boundaries between these fields of study. With regard specifically to our project, it also helps that Dewey's approach assumes interrelation between thinking and doing and also a similar interrelation between education and wider social phenomena. As we wrote and edited this book we frequently stopped and asked each other whether we were talking about a particular thinker's philosophy of mathematics or their philosophy of mathematics education. Occasionally, we also questioned whether we were talking about "doing" philosophy or actually doing math. With Dewey, we realized that these distinctions were not particularly meaningful, as his approach downplays the worth of thinking about any of these phenomena independent of the others. Hence, in DME *philosophy of mathematics is not separate from mathematics, and neither are separate from school.* They interrelate and influence each other. As we established in the previous chapter, these interrelations have the benefit of cultivating agency in student mathematicians. Agency isn't limited to applying pre-formed mathematics to possibly pre-established problems. Instead, what mathematics is and how it should be used given our social needs are all up for scrutiny and negotiation. We see this agency as important for the development of individuals, for the enrichment of mathematics as an enterprise, and for the flourishing of our democracy.

DME and the Case for Outsight

In Chapter 1, we spoke of Toulmin's notion of outsight and how it permits us to think philosophically about mathematics from the position of the interested outsider. From the "outside" it is easier to see that broadening our understanding of what is to be considered mathematics will be beneficial for mathematics education. The recognition that human interactivity with our physical and social environments is central to mathematics can help math class become a place where students do not simply learn about mathematics or construct idiosyncratic systems that have no meaning outside insular structures, but where students can be empowered to mathematize as a means to identify and creatively confront genuine problems. Furthermore, when we broaden our understanding of mathematics—or at the very least when we pause to take stock of our current beliefs about a subject whose prominence we may never have questioned before—we are in a position to reconsider its affordances and limitations.

Here, near the end of our project, we are compelled to say more about who we are, about our own mathematics autobiographies, and about the paths that led us—both separately and in partnership—to pursue a philosophy of mathematics and of math education that more earnestly and effectively supports democratic aims. Though each of us has experience teaching math to young students, Kurt in middle school and Cat in elementary, we each have a different sense of belonging in the world of mathematics. Kurt would describe himself as more of a mathematical outsider, having landed somewhat accidentally in the lap of mathematics (while also having had a lifelong affinity for it). His interest in school mathematics gave shape to his scholarship in the philosophy of education, a field in which he is now an insider. Alternatively, Cat had more experience with formal mathematics content, having also served for some years as an instructional coach and math methods instructor, and instead seemed to fall into philosophy. Though our paths were transposed, we each came to recognize the critical role of both domains, of school mathematics and philosophy of education, in the human drama. Our partnership is, we feel, an example of how different experiences and perspectives serve to enrich an enterprise such as this. The outsight people bring to a project is as critical as any expert insight, and math education is well served by the voices and contributions of many, including professional mathematicians, students, educators, caregivers, and philosophers. It is not just that they are among the many stakeholders who deserve vibrant and meaningful mathematical education. Mathematics education is in need of *them*, of their insight and outsight, to reach its full potential as a democratizing force.

Taking our cues from Dewey, we know that there is no end to the quest to be democratic; there is no destination to reach or summit to scale. Democracy is an ongoing project. What we have put forth in these pages is our attempt to contribute to it by thinking about how school mathematics can be part of the democratic process, given its current state. We freely acknowledge that our practical suggestions for math class are just some of the many viable ways for mathematics education to be democratic. What we do see as more foundational is our argument that we need to think differently about mathematics in order for its teaching and learning to be able to contribute meaningfully to our social needs and to life in general. It is likely that there are other philosophies of mathematics education that could serve as a support for the mathematics classes that we need. What we have done in this book is develop one suitable philosophical approach to mathematics and mathematics education specifically to do so. In keeping with our pragmatic approach, it is not intended to be a philosophy of mathematics education for *all* time, but since it has been developed with the problems of the day in mind, it is a philosophy of mathematics education for *our* time.

Note

1 The inward turning to which we refer can be either to the individual mind (for psychological constructivists such as Piaget or von Glasersfeld) or the social or cultural group (for the social constructivists such as Ernest).

References

Andersson, A., & Österling, L. (2019). Democratic actions in school mathematics and the dilemma of conflicting values. In P. Clarkson, W.T. Seah, & J. Pang (Eds.), *Values and valuing in mathematics education: Scanning and scoping the territory* (pp. 69–88). New York: Springer Open.

Ernest, P. (2002). Empowerment in mathematics education. *Philosophy of Mathematics Education Journal*, 15. http://socialsciences.exeter.ac.uk/education/research/centres/stem/publications/pmej/pome15/ernest_empowerment.pdf

Frankenstein, M. (1983). Critical mathematics education: An application of Paulo Freire's epistemology. *Journal of Education*, 165(4), 315–339.

Glaude, E.S., Jr. (2007). *In a shade of blue: Pragmatism and the politics of Black America*. Chicago: The University of Chicago Press.

Nakawa, N. (2019). Mathematical values through personal and social values: A number activity in a Japanese kindergarten. In P. Clarkson, W.T. Seah, & J. Pang (Eds.), *Values and valuing in mathematics education: Scanning and scoping the territory* (pp. 157–170). New York: Springer Open.

O'Neil, C. (2017). *Weapons of math destruction: How big data increases inequality and threatens democracy*. New York: Broadway Books.

Orrill, R. (2001). Mathematics, numeracy, and democracy. In L. Steen (Ed.), *Mathematics and democracy: The case for quantitative literacy* (pp. xiii–xx). National Council on Education and the Disciplines. Princeton, NJ: Woodrow Wilson Foundation.

INDEX

Page numbers in **bold** refer to tables